LIFE · IN · THE

OIL FIELDS

LIFE · IN · THE
OIL FIELDS

By Roger M. Olien and Diana Davids Olien

★

TexasMonthlyPress

Texas Monthly Press, Inc.
P.O. Box 1569
Austin, Texas 78767

A B C D E F G H

Library of Congress Cataloging-in-Publication Data
Olien, Roger M., 1938–
    Life in the oil fields.
    Includes index.
    1. Texas—Social life and customs.   2. Petroleum
industry and trade—Texas—History—20th century.
3. Oil industry workers—Texas—Interviews.   I. Olien,
Diana Davids, 1943–
F391.046   1986   976.4′06   85-28828
ISBN 0-87719-029-1
Book design by Mark Grasso

For
Christina

Contents

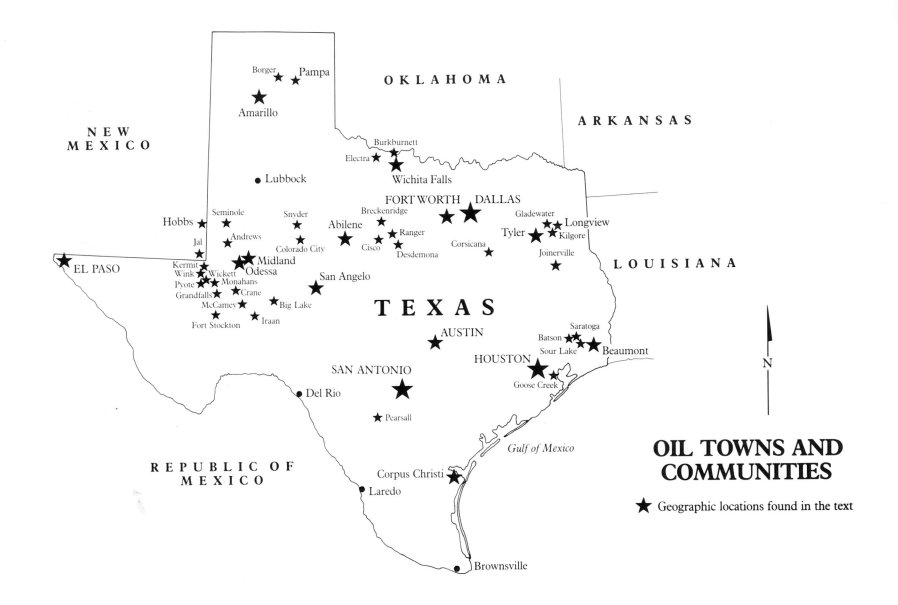

OIL TOWNS AND
COMMUNITIES

★ Geographic locations found in the text

Preface

We are now more than a generation removed from the oil patch as it was. Wooden rigs, ragtowns, and Choc beer are now only fragments of a way of life that is being forgotten, even in the very places where it had great impact. As Texas moves hopefully into the postindustrial age, it is easy to forget that oil changed Texas from a rural and agrarian state, with slender hope for economic power that would match its physical size, into a large and wealthy urban state. In one sense, a second Texas Revolution began at Spindletop on January 10, 1901. As it ran its course, it brought Texas into the mainstream of modern American life. It also brought social and economic changes that still define the character of the Lone Star State. The shadow of the derrick lies long over modern Texas.

It is easy to forget that life in the oil fields was rough, tough, and anything but glamorous. Its industrial setting has had little appeal for nostalgic romantics. For that reason, most writers have described only the sensational parts of the scene: the giant gushers, the wild blowouts, the raucous boomtowns, and the gaudy self-made millionaires. They have all seemed more interesting than ordinary people and everyday life.

We have chosen a different approach: the use of oral history to reconstruct the regular and common parts of a bygone way of life.* There are some references to events in Oklahoma and New Mexico, but our geographical focus is Texas. In some instances, in order to make recollections easier to follow, we have condensed statements, omitted unnecessary *and*'s, *you know*'s, and similar terms, or added a word or phrase in brackets to clarify a statement. We have tried, however, to stay as close to original expressions as coherence permits; we have not changed grammar, figures of speech, or salty phrases.

We owe special thanks to the hundreds of interviewees who shared their recollections with us and with other scholars over the years; theirs is the greatest contribution to this history. Our work was also greatly aided by Dr. Bobby Weaver of the Panhandle-Plains Museum in Canyon, Texas; Mrs. Betty Orbeck of the Permian Basin Petroleum Museum in Midland; Dr. Joe L. White of the East Texas Oil Museum in Kilgore; the staff of the Barker Texas History Center at the University of Texas at Austin; and the depositories of the photographs we have used. At the University of Texas of the Permian Basin, Dr. Duane L. Leach,

Dr. H. Warren Gardner, Dr. Mary E. Quinlivan, Bonnie Holly, Jean Tims, and Martha Edwards offered special help and encouragement; Gayla Carbajal was responsible for the expeditious and accurate typing of the manuscript. Leo and Betty Byerley and J. Conrad and Kathlyn Dunagan gave the manuscript patient and careful readings at various times. Scott Lubeck and Chris Kindschi of Texas Monthly Press gave us helpful advice and welcome editorial assistance. The interest and aid of all of these people should not be taken to represent full agreement with our opinions or those of the interviewees we have quoted.

Midland, Texas
March 1985

---

*We have not described oil field work because this is the subject of a separate book, now in progress.

Introduction

For almost a hundred years, from Edwin L. Drake's gusher of 1859 to the middle of the twentieth century, the discovery of oil led millions of Americans to work and live in the oil fields. As exploration and production spread from Pennsylvania to the adjoining states of New York, Ohio, and West Virginia, and from there to the Midwest, Southwest, and West, a vast human tide of workers, contractors, businessmen, and their families engulfed successive parts of the United States. Where this tide reached, country villages grew into cities, and towns appeared on empty plains. When the tide ebbed, it carried with it farm boys and small-town people eager to seek their fortunes at its next destination; left behind were those who produced the oil found as well as those who were ready and able to settle down. This great migratory tide shaped the lives of those who joined it and transformed the regions through which it moved. Like all great events, there was more to this one than appeared on the surface. It was more than a migration from oil field to oil field. When the wave reached the American Southwest, it was a way of passage from rural farm life to urban industrial society for hundreds of thousands of Texans.

They came to the oil fields in wagons, on horseback, by train, and on foot. They came in Model T Fords piled high with tents, footlockers, and children. Between 1901 and the most recent oil boom, generations of Texans came to the oil fields. They left the hardscrabble farms and drought-parched ranches in search of opportunity and excitement. The oil fields promised both.

In Texas as in other states, the main lure of the oil field was economic opportunity. Like the Yankee farm girls who went to work in the Lowell textile mills, like the French Canadian villagers who tended looms at Amoskeag, like the rural Eastern European emigrants who found work in the sweatshops of New York City, the Texans faced the hazards and challenges of a new life because they saw the promise of a better one, for themselves and for their children. The oil fields offered a chance to get ahead—or, at the very least, an escape from the soul-numbing drudgery of being dirt poor. For the dreamers, there was always the hope of striking it rich. Everyone in Texas had heard alluring tales of instant wealth and overnight millionaires. Perhaps few of the farm boys expected windfall riches, but they could hope for regular pay.

The oil fields offered attractions apart from money. Exploration, with elements of high risk and high stakes, with fortunes made and lost overnight, was more exciting than the plodding routines of country life. It was fast-moving, unpredictable, and constantly challenging. To do oil field work and to cope with oil field life, it was necessary to be healthy, strong, and adaptable; for that reason, the oil field population was young. The farm boy who went to the oil fields found himself working with other farm boys, who were as ready as he to work hard and play later. There was nothing quite like it back home.

Whether workers entered oil field life in the steamy swamps of the Gulf Coast, the parched Permian Basin of West Texas and southeastern New Mexico, or the wind-scoured plains of the Texas Panhandle, before 1950 they found that life was much the same from field to field. Wherever they went, life and work took a distinctive form because they were keyed to the economics and technology of the oil industry. The sequence of events in one field was much like that of the others. A large discovery ordinarily led to rapid development, which created an intense local demand for labor. As each producer tried to extract oil before producers on adjoining leases got it, drilling went on at a breakneck pace, slowing down only when a succession of dry holes marked the limits of the newly discovered field. During the short period of frantic drilling and all-out production, labor was in short supply, jobs were plentiful, and wages were top dollar. Seasoned workers and green farmhands alike rushed to new discoveries while there was a pressing need for hands to prepare drilling sites, to build rigs and tanks, to drill wells, and to lay pipe.

A new oil field created as great a demand for those who provided basic services—nailing together jerry-built housing, cooking steak and eggs, renting cots, selling work clothes, repairing cars and trucks, and showing motion pictures. And there was no lack of others who catered to different needs—for cold beer, a fast-moving card or dice game, or easy-going female companionship. All these people joined those who worked out in the field in flocking to the scene of a discovery. They made the boom.

As oil field veterans knew, boom usually turned to bust. Once a field was defined, once leases were drilled up and tanks and pipelines completed, hiring dropped sharply. A few workers who wanted to stay on kept jobs as pumpers and pipeline employees; a few settled down in small-town business. Some workers dropped out of the shuffle from field to field and took their earnings back to the farm or

A farm boy's first view of the oil
patch was a forest of derricks like
this one near Beaumont, early in the
twentieth century.

sought work in the growing industrial centers of Texas, Oklahoma, and Louisiana. Most people packed their suitcases and moved on to the next boom. They always expected to do so.

This was the pattern of oil industry activity in the field before 1950. Behind this pattern was the volatile and fragile economics of petroleum. Wide swings in the prices paid for crude oil marked its history. High prices, or the expectation of them, prompted exploration that often led to the discovery of major fields. Once those fields went into full-scale production, however, prices usually fell. Exploration then slumped until prices rose or until an exciting geological opportunity, like that in East Texas in 1930, stampeded oilmen eager to seize it and created a new boom. Hard times followed fast on the heels of flush times. Change was constant and unrelenting, though not exactly predictable. This meant that making a living required moving on: geology and economics determined where and when.

Industry economics determined the common features of oil field life. Everywhere they went, most people were part of a moving crowd. They could expect to live in makeshift quarters that were often uncomfortable and sometimes unsanitary. They had to do without possessions they could not cram into a suitcase or tie onto a car. Their children were newcomers in crowded schools; they were newcomers in communities not always cordial to outsiders. Far from home and kin, they had to be self-reliant in family crisis or fall back on the help of those who moved as they did. From time to time, they could expect to be out of work. The instability of industry economics created these disagreeable features and gave oil field life its dimension of impermanence. Expectations were always in flux. Perhaps the next boom would mean better housing; perhaps it would bring the chance to do well and settle down. The present was always a passing stage on the way to something new and different.

If the constant mobility of most oil field workers was the result of industry economics, some of the ruggedness of oil field life was the result of geography. A few large oil fields have been discovered in or near large cities—Long Beach, California, and Oklahoma City, Oklahoma, are examples—but as geological chance has had it, most oil has been discovered in remote, thinly settled places, in swamps and mountains and deserts often accurately described as godforsaken. More often than not, living in the oil field meant roughing it in relative isolation from large commercial centers; it meant being part of industrial life in rural, or even

The cleaned-up drilling crew and
bystanders at the discovery well of
the Hendrick field in 1927.

BRINGING IN THE WESTBROOK & CO. WELL Nº 1
ON T. G. HENDRICKS RANCH, WINKLER COUNTY, JULY 22, 1926.

wilderness, surroundings. No wonder country folk went to the oil field in larger numbers than their city cousins: they could get by without indoor plumbing, central heating, and a grocery store around the corner.

Country people were also used to rugged, outdoor work done with lots of muscle and little book learning, and that was what they found in the oil field. Jobs in the oil field were as rough as living conditions. Most work involved long hours at hard and occasionally dangerous tasks, done in all kinds of weather. Before World War II, the prime requirement for most jobs was strength and endurance. Rigs were built by hand with saws and hammers; ditches were dug with pickaxes and shovels; drill pipe was assembled and broken with tongs and rope. Hard work was followed by hard play. As in nineteenth-century cattle towns and mining camps, the fun and pastimes of oil boomtowns appealed to restless young men looking for a good time. Oil boomtowns were rowdy around the clock; middle-aged and middle-class people were usually shocked by what went on in them.

Oil field towns can still be rowdy on occasion, but since World War II, life and work in the oil field has changed in many ways. The single greatest change has been the shrinking of the once-vast tide of people in constant motion from field to field to a small stream of restless, rootless persons. This radical departure from earlier times came about for a variety of reasons, including changes in the economics of the industry. After World War II, American oil companies began to spend more dollars looking for oil overseas than at home; there were longer periods between major discoveries, and the pace of exploration stayed slow. From the late fifties to the early seventies, workers had little action to follow. They settled down. Many drifted into more secure and less physically demanding jobs. Fewer new workers entered the industry. The opportunities and excitement were gone.

When domestic exploration picked up in the early seventies, the floating work force of eager farm boys that had sustained the oil booms of earlier decades had disappeared. Farm boys were a vanishing breed, largely because there were ever fewer farm families. With the steady depopulation of rural America the oil industry lost its main source of unskilled labor. Replacing the country boys were the town-raised children and grandchildren of oil field workers and, in Texas, Mexican Americans, and an increasing number of men from desperately poor rural Mexico. In little more than two decades, the accents of many oil field work crews changed from Texas twang to Mexican Spanish; an entirely new cultural strain entered the oil field.

The familiar Texas panorama: oil derricks scattered through boom-time Corsicana in 1898.

When exploration was slow in the sixties, the workers who stayed with oil gave up the nomadic life for the settled enjoyment of material prosperity in oil field towns. They were able to do so for a number of reasons. The extensive postwar road building program in Texas broke the isolation of rural communities; with the obvious exception of off-shore exploration, even remote oil fields had paved roads linking them to population centers. With the work site no more than a two-hour drive away, workers lived in town and commuted to work. When automation in oil production put operations once monitored by workers under the control of electronic systems, workers who once had to live on the job left isolated leases and stations for life in town. In all, fewer and fewer workers lived out in the field, within walking distance of work. Moving to town and settling down ended the physical and emotional isolation of their wives and children; it allowed them to become active participants in communities they once moved through. Where workers once did industrial work in rural surroundings, they now could make the final step in the transition from rural to urban industrial life by moving to the towns and cities.

Because of these changes, life and work in the oil field today is different from what it was before 1950. It has become more like other types of urban life and industrial work, closer to the mainstream of American life. Customs, attitudes, and values generated by past oil field experience, however, have survived the changes and extend far outside the oil patch.

Some of these customs, attitudes, and values came from the country and were modified in oil field surroundings. The country custom of helping neighbors and kin in time of need, for example, became part of oil field practice as oil field people on the move helped one another in the absence of local friends or relatives. It survives today even in city neighborhoods in places like Midland and Kilgore. The respect for "macho" behavior, physical strength, and endurance is another survival from rural life. Physical strength was as necessary to the rig builder or pipeliner as to the farm boy or cowboy, and, carried into an urban setting, physical prowess and aggressiveness continues to be glorified in Texas high school football.

Other values and attitudes may have come from the country but gained particular strength in the oil field. The premium placed on personal independence and self-reliance may have had rural roots, but in the mobile world of the oil field, being able to survive on one's own was a necessity. The belief that ambition and hard work usually lead to financial success waned on the submarginal farms of Texas, revived in the oil fields, and lives on in most segments of Texas society, because it proved to be true. Many poor farm boys made a modestly prosperous living in the oil industry, and that measure of success reinforced their belief in getting ahead through hard work. Belief in economic independence and opportunity have been translated, in recent decades, into contempt for handouts, dislike for the welfare state, and disdain of welfare recipients.

The high value placed on personal independence, taken with the working conditions common in the oil field, also influenced attitudes toward work in the oil patch. Early oil field workers, many new to industrial life, did not hurry to accept industrial discipline. Some of them drank on the job, took days off when they felt like it, "mouthed off" to the boss, and took being fired as a matter of course. In a work world where few jobs lasted long and a better job might come along tomorrow in the next field, such behavior was neither unusual nor a particular liability. Oil field workers were commonly no more eager to accept the discipline of labor unions, and their attitudes, as well as their mobility, made them hard to organize. Today oil field workers remain one of the least organized labor groups and suspicion of unions can often be found among employees as well as employers.

The economic uncertainties of oil field life—boom to bust, in and out of work—together with the realities of oil exploration, in which there are dry holes along the road to success, meant that oil field residents had to accept taking risks and facing uncertainties. Making a virtue of necessity, they put a premium on daring and risk-taking, in pursuit of their visions of a better life. These visions were often realized, at least in part. By taking a chance and leaving the farm, many of the workers and their families were able to enter the mainstream of American consumer society: even if they did not strike it rich, many of them advanced from poverty to relative comfort. In varying measure, they confirmed the American dream, and they often became its most determined defenders.

The political conservatism of the oil patch, like so many other aspects of the life and culture of modern Texas, can be understood only in the context of the customs, attitudes, and values that were preserved and transformed in the oil fields. The shadow of the derrick reached into every corner of the state as oil activity spread across it and as oil field workers settled into more permanent jobs in Houston, Dallas, Fort Worth, and the other cities of the state. In one sense, the diminishing reserves of oil that remain under the surface of the land may be less important than the surviving culture of the oil patch. To paraphrase a popular saying, you can take a Texan out of the oil field, but you can't take the oil field out of the Texan.

Work and a chance at greater fortune
drew farmboys to the derricks and
slush pits of the oil patch.

1: Making the Scene

Well, I drove off down there one day. There was a couple of fellows there worked on the crew. I said, "Why are you setting down?" They said, "Well, Mr. Joiner's off trying to find a testing tool, get somebody to come up here and make a drill stem test." I said, "Oh, you don't mean it!" They said, "Yes, sir, he's supposed to be back tonight or in the morning." I thought, "Well, I'm just wasting my time," because from everything you heard, it'd been salted since it was fifteen hundred feet deep.

I went back the next day. Sure enough, they had a testing tool. It was in the afternoon. I got there, and they had trouble getting in the hole. They'd had to stop and circulate and whatnot. They got in the hole just about sundown, just dusk. They dropped that in there, and man, here comes gas and oil, gurgling and going on. And about that time here comes the oil, over the derrick, and all the pine trees downwind from us for several hundred yards were just soaked with oil.

That gusher described by oil scout **Paul Davis**, from the discovery well of the giant East Texas oil field, began the greatest oil boom Texas has ever known. It followed the first, at Corsicana, by thirty-five years; the last boom has yet to happen. People, equipment, and technology changed over the years, but time never dulled the keen excitement following the discovery of a new oil field. The torrent of oil over the top of the derrick was like the starting gun in a race. Brokers, promoters, workers, and thrill seekers raced to the scene, all hoping for overnight fortune, all gambling on the future. Some risked their money, some risked their lives, but all were caught up in the whirlwind of action set in motion by the discovery of oil.

The first workers to arrive in the human stampede to a new field were the boomers. Like those workers who followed railroad booms of an earlier day, the boomers wanted more than a job: they wanted to be with their friends, where the action was hottest. Several veterans of Batson, Sour Lake, and Goose Creek recalled the boomers of the earliest years of Texas oil.

W. H. "Bill" Bryant: A boomer is like this. You get into a field; when it goes dead, the boys drift off, and you feel no good. You know all your friends are over there in that field, and you know there are a bunch of other guys over there. It's like going to the show. This field is dead here. There's no fistfightin', and drinkin's kind of slack, and not much gabbin'. Town kinda civilized.

Well, you go over there, and probably you get in there and start running a rig. Boy, when you get in there with a good rig, with probably thirty or forty rigs right around you, and everyone trying to beat the other one, it's fascinating. All of them have whistles on there, and when you start off there, all them whistles blow, trying to beat the other guy out of the hole.

Excited and satisfied, workers watch Salazar B-2 flow in Duval County.

Early C. Deane: We—a lot of us—wouldn't stay after the wells go to pumping. It didn't look exciting enough for us. We'd hear of a gusher being brought in at Sour Lake and then after that Batson, we'd move on. Those that stayed and kept their property or acquired some, just produced oil from pumping them, they made money. They were the ones that stayed. We were boomers, I guess. We'd go to Batson, and Saratoga, and to Jennings, Louisiana. We'd like to see them gush.

I mean that we liked it where it was more exciting, big new fields would come in. And the same people we knowed, they'd say, "Yeah, got a job for you at Batson." We'd go there. Get sort of tired of this business of watching old pumping wells, and greasing up the engines, and looking after the stuff, and gauging the tanks. We'd rather go to where the big doing's on, gang up there again.

Frank Redman: A boomer was the best mechanics there were. See, they traveled from shop to shop and learned all the tricks of the trade.

Boilermakers, machinists, tinners, you know. And they'd go ride freight trains from one division to another. Work till payday and then go to another. Always green grass ahead, you know.

One boomer, *O. C. Profitt*, traveled through North and West Texas with his family:

I've been drilling sixty-three years. I've drilled everyplace. I started in Hamilton County, Texas. Started working in 1915 but started drilling in '16. I was raised on a farm; went to school, little country school. I was the first Profitt, I guess, that ever worked in the oil fields.

I pumped water for a little bit, and then I went to helping them on casing and dressed tools. Then, I was a smart boy: I went to drilling. And from then on, I went to Ranger and Hogtown. And I was using a cable tool rig. I learned to build rigs, too. Learned to do every part of it.

I moved around a lot. From boom to boom. We'd be getting top wages in one boom, and there'd be somebody drilling a well, and I'd hear of the boys going up there. I'd go up there and go to work for the same money and worse living conditions.

Wages weren't necessarily better in the booming areas. I'd go to the booms just in order to be with the action. I just liked to be where the action was. And guys—I never got into scrapes myself, but I liked to go watch them guys get drunk and somebody get whipped real good. I guess you'd call me a boomer.

My wife and I moved around. We lived in Fort Worth several times. I'd leave Fort Worth and come to West Texas. I liked to work in West Texas. Soon as I'd picked up a little grocery stake, why, I'd go get 'em and bring 'em out. I worked real good, real good. Been living with the same woman fifty-five years. Had our fiftieth wedding anniversary about five years ago. We managed to make some money. That's the important part.

Not all workers followed oil for thrills. Some preferred to work for short periods and move on when it suited them; the intermittent employment usual in oil field work fit their lifestyle. Transients and drifters, working for many bosses and answering few questions, dug many a ditch and laid many a pipeline, as these oil field veterans recalled of the thirties.

Clell Reed: Oil field tramps, they'd come in. In those days the connections foreman had the right to hire you right on the spot. No examinations, no anything, just go to work. The timekeeper got your name: no Social Security number, no nothing. He kept around fifteen hundred to two thousand dollars in his billfold. Payday, quite a few of them would decide to quit. They called it "dragged up"—going to drag up, quit. When the job's finished, that's "flanged up." Flanged up. Anyway, those men the timekeeper paid off in cash. Just got a little receipt. Then they'd crawl under the fence and go to work for some other company. They were just drifters,

Boomers flood Kilgore in 1930. The crowd on the right is in line at the railroad depot for passenger tickets.

tramps. But those days, if he got fired the fore-man'd say, "Go to the timekeeper and get your pay, 'cause I just drug you up."

W. H. Bond: Back in those days, during the Depression, there was more labor. They just come there. A certain type of people followed that work. I'd say sixty, maybe seventy-five, per-cent of them were not married. They just fol-lowed that trade. They were not married, most of 'em. They stayed in those pipeline camps.

L. E. Windham: You take up in the middle thirties when they would—say Humble or Shell was going to lay a line, you would see those old boomers come in. They would be in two or three weeks before she started. They would come in a few days before the work started. We had three or four family-style dining rooms in McCamey. Well, that's where all of us ate. You could get a meal for thirty-five cents or three meals for a dollar. They would all be lined up on the way there, and they would be mooching you for meals or anything. They were on canned heat

and everything. They all had nicknames. You would see them on the roads coming in. They would make that payday and when that work was over, they were gone. They were on the tramp the rest of the time, mostly.

Many workers entered oil field life because it was the only well-paid work available. After both world wars, for ex-ample, veterans looked for oil field jobs after their hitches in the armed forces ended. *C. J. Davidson* returned home to San Antonio after he was discharged from the Navy in 1919:

There wasn't anything much to do around there but pick cotton, and the crop was kinda short that year, so I decided I had better go to the oil fields. Burkburnett was having a little boom, and Ranger and Desdemona, and there was some activity down around Corsicana. So in connection with a friend of mine who was in the Navy with me, we decided to go to the oil fields. We came to Fort Worth from San Antonio and stayed around a few days until our money got short, and Sam went back home. I decided that I was going to Desdemona. I had my choice, to

Crews like this one, photographed on a rig near Corsicana in 1911, followed work from the Panhandle to the Gulf and from the Permian Basin to East Texas.

go to Desdemona, Ranger, or Electra, Burkburnett, etc., but Desdemona was a little closer to Fort Worth, and my money was short, so I went to Desdemona.

In Desdemona I did some roustabout work and helped a surveyor lay out pipeline right-of-ways, just any kind of work I could find to do. Things got a little dull in Desdemona in late 1920, and I migrated up to the Electra-Vernon area. Things were getting a little active there. Electra was an old field, but it was getting more active and it spread over into Wilbarger County and other areas. After working up there on a well or two, Breckenridge was beginning to pick up. I worked on wells, pipeline work, and whatever I could find to do around Breckenridge.

In 1920 things were so tough around there I got a job in a bank, more or less as a janitor, but I got advanced a little later and got a little more pay out of it. But I knew I wasn't cut out for banking, so I continued to work in the oil fields in Stephens, Young, and Eastland counties,

dressing tools, working on casing crews, whatever there was to do.

Large numbers of farm boys, used to working long hours at hard outdoor work, found the oil field a big improvement over the farm. ***William H. "Bill" Measures*** left the family farm for the oil field in the 1920s:

> I grew up on a farm out in Weatherford, Texas. My father was a water well contractor. My mother lived on the farm, but Dad made a living drilling water wells. We raised what we ate, and that's about what you had.
>
> A friend asked me if I'd like to go out and work in the oil field. I went out there, and he had a man named Ernest Mason that was a dirt contractor. So I went to work.
>
> They fed like a chuck wagon, like they did on a ranch, and just camped out. So we went out and told the old boy we was looking for a job. And he said, "Well, you better stay and eat dinner. Sit down." Boy, they had real good food. The cook was real good.

Roughnecks, roustabouts, and teamsters in Wink in 1928.

The boss said, "Can you drive a tractor?" I said, "Sure." He said, "Well, get on that." We had a little old Fordson tractor on the farm and an old steam threshing engine, that's all I'd ever seen. I got on that tractor, and, boy, I never got off of it till I went back to school. We was building levees and things like that. Well, that tractor would do the work of twelve big horses.

Clell Reed left the farm in 1930:

I was raised in the cotton patch down in Mills County. I had made a crop of fifty acres and one bale of cotton, and he [his father] said, "Son, you'd better go to the oil field like your three brothers." They were in the oil business, working for oil companies. So I had a football sweater and a pasteboard suitcase, about two or three dollars in my pocket, and I set off. I hitchhiked to McCamey in 1930, in March. Got there about dark. It was sleeting, and just cold, cold. And I was standing at a Humble service station— my brother had helped me for an interview out there—so they met me there. I was brought to Midland the next day, and I worked for the [Humble] sales department for about three years.

Friends and relatives often encouraged young men to join them in the oil fields, where they arranged jobs for them. **Hood May** recalled his own start:

How did I come to do oil work? I really don't know. I was just looking for a job and got up in Ardmore, Oklahoma. I had a friend who was running a casing crew, running pipe. I knowed him at Hoffman, Texas, where we lived. In fact, he and I had gone to school together, and he was the one got me started in the oil business. I never would have went out there wasn't for him, 'cause I didn't know a thing in the world about it. He got me a place to stay there; I was pretty broke. I guess the third night I stayed there, he said, "We're going to run some pipe. I want you to go with me." I said, "I don't know anything about running pipe; this is the first oil field I was ever in." "Oh," he said, "come on. I'll teach you."

Well, I went out with him. We run a string of eight-inch, and I think it was about sixteen hundred feet of it. When I came back, I was pretty tired because I wasn't used to that kind of work. And if you're not used to it, you work harder at it 'cause you don't know what you're doing. I come in and slept awhile, and he come and called me again, said: "We're going out on three strings of big pipe." "Well," I told him, "I don't want to go. That's too hard work." "Oh," he said, "you can make it."

I went out there and made that three strings of big pipe and got pretty good money for it. I'd come in, and I was so tired I couldn't hardly sit down. I was sore and everything. He come around and paid me and I said, "That's pretty good money. I believe I'll just stay with it."

A friend persuaded **W. E. Mapp** to leave the farm and go to the oil field after World War I:

I made the job in January of 19-and-20. Desdemona, Texas: Hogtown. I was raised in Mississippi. A friend of mine had been out here and came back. He had just gotten out of the Army and came out here and worked awhile at Albany. Came back home. And then he came back, and I came with him. 'Course he knew the jobs.

I got into a crew, was all of them singles, and we run all over the country. So I guess I worked in and out of Desdemona there for six to eight months. Then shifted to Rising Star, Texas, and I was down there for a few months. I left the oil fields for about eighteen months. I worked in a sawmill [in Mississippi] for a while, and I worked in a planing mill. Then, when I went to California, I worked for the Southern Pacific Railroad, for a while, and for Southern California Edison Company. Then I came back to Texas, I guess it was in '24, putting tanks together.

Even without the help of relatives and friends, it was not hard to find work in a booming oil field. **W. H. "Steamboat" Fulton** got a job drilling near Mineral Wells in 1919:

We got in by train at about six o'clock in the evening. I got a taxi driver, and I said, "Do you know any place where we'd get meals and room

Hours on the job were long: the
night crew sits in front of the day
crew on Fair No. 2, near Wright City
in East Texas, in the 1930s.

for a few days till I get located?" "Yeah," he said, "I've got a friend that's running a rooming house."

He took me over there, and they put me in a room right off the dining room. While we was sitting there in the room, we could hear the men out in the dining room talking. I said to my wife, I said, "Them guys is drillers, tool dressers, out there." We got cleaned up, went out on the veranda. I said to the lady that waited on the table, I said, "Are those men all drillers?" And she said, "They're working out there in the oil field." South of Mineral Wells.

So we finished eating our dinner. I went out and set down beside a guy, and I said, "Do you guys—is there any drilling going on real handy?"

"Yeah," he said, "there is a boom out here south of town." He said, "Are you a driller?"

I said, "Yes, I am."

He said, "Are you looking for a job?"

I said, "That's what I come out here for."

He said, "The tool pusher is uptown hunting for a driller to go out tonight at midnight."

I said, "I've been riding the train and haven't slept any for five days. I wouldn't be fit to go out tonight midnight and stay awake."

He says, "C'mon, we'll walk up the street, and see if we can find him."

And we started up the street, and we met him. And that guy hired me in one minute to go out at midnight.

Clarence Dunaway remembered hiring a hand for his rig building crew in Odessa in the thirties:

He was Blackie. He was half Indian. He's just an old Okie, is what he was. Great big old strapping boy. I guess he weighed 180 pounds.

He come in there. He didn't have no crown in his old straw hat, and his shoes had pasteboard in them. Shirt was tore.

We needed a hand that morning. I went out there in front of the cafe. I said, "Did you ever do any work in the oil field?"

"Well, I drove a truck at Wewoka."

I said, "Was you ever in a derrick?"

He said, "Well, I helped them put up the first section lots of times."

I said, "Well, you want to try to go out and help us?"

"Yeah."

I said, "You had any breakfast?"

"No." And he said he hadn't had no supper either.

Young men like Carl B. King (later a prominent Dallas oilman) apprenticed as tool dressers on cable tool rigs and often advanced to become drillers.

Well, I took him in there in that cafe and told them, "Give him something to eat and a lunch." And I had an old big boy working for me named Oliver Wheat; he said, "I got a pair of shoes up to the house, I'll go and get him a pair of shoes and a pair of overalls."

And in a week's time, that son of a gun was showing them a new way to go! Oh, man, he made us a good hand! Yeah, he was big, stout, and young. Man, he just throwed them scaffolds up!

But they killed him over there in Normandy.

Most oil field jobs were as easy to lose as they were to get: for oil field workers being part of the "suitcase parade," being in and out of a job, was a way of life, as *Clell Reed* remembered:

They'd hire a bunch of men to lay a big line. They'd take 150 men, maybe 200, to lay a big ten-, twenty-inch line. When it was over with, that's when the suitcase parade came in. And there was no labor union; they didn't have to go

before a board to let them know why they let everybody go. When they'd start another one— here comes the suitcase parade coming in.

V. L. Cox recalled similar conditions in rig building in the thirties:

At that time you'd hire out to one contractor, and if you was lucky, you'd work five or six days. They wouldn't have another job. You'd hire out to anybody that wanted a man. That's what I liked about it. You'd go fishing anytime you wanted to, didn't nobody care. They'd hire another man in your place. No trouble at all.

Bosses were quick to fire workers who didn't suit them.

Hoke Tehee: One time we were working for old W. E. Skelly. They was laying a pipeline. So he come through, and them boys, they was kind of goldbricking. So he just canned the whole bunch. So I happened not to be there. He missed me, you know. So I just kept on working for him. Yeah, he was pretty strict with us.

Don Dittman: Old Checkbook Myers, know how he got that name? He'd fire men off the job. Carried his checkbook with him, and he'd pay 'em off on the spot if he didn't like the way they were working. I had a reputation for being just about as bad. They always said I was rough to work for. I just expected 'em to do what I'd do. If they didn't work, I'd get somebody else.

Orville "Checkbook" Myers: The driller was pulling the bailer out. It caught under the pipe somehow or another. It broke the line and jerked the tail post out on the sand line, and it hurt the driller's leg. I took him to the hospital.

The tool dresser just set down on the bailer's bench and set there instead of doing anything, trying to get the tail post put back in. And when I come back and he hadn't done anything, why, I fired him. The morning-tour [pronounced "tower"] driller who was working the well, he gave me that name of Checkbook. That's how I got the name. I paid him off at the time. I just come back and asked him why he hadn't done anything. I don't remember what he said, but

anyway he hadn't done anything, so that's how I come to fire him.

Bill Sidwell told me when I come down here to push tools—I told him, "I don't know anything about pushing tools"—he said, "I want to tell you one thing that I want you to always remember: it doesn't cost you any more to pay a man off today than it does payday." And I tried to remember that, and that's the way I run the business.

As husbands and fathers moved to keep working, wives and children moved with them, in order to keep families together. *Bessie Leonard* remembered how this was done in the 1920s:

When you moved, that was really a deal. You just tied stuff all over your car. The thing of it was, them days, it didn't seem like anyone had any more than the other guy. We all lived in about the same standard of living. They were looking for work. They were migrating from back in the Eastern states. Some of my best friends that I had at that time came from back

East. And they had no more than we had. They would just load their car with what they could put inside, on top of it, on the running boards—they had running boards at that time. And they'd just tie the spare tires on somewhere and take off. Wherever they got a job, that's what happened.

We came through here [West Texas] in a Ford roadster, and what we could haul in it. They had isinglass on the sides, if you were lucky, and little curtains you could put up. And that was it. And nobody lived any better than the other guy.

We didn't mind it too bad. We just didn't really pay that much attention to it, I guess. It was kind of a lark in a way.

Mrs. R. V. Wilson: I arrived in Crane County in March 1927. I made the trip from Levelland with a doctor's wife, Mrs. Richardson, and a girl whose father had the bakery in Levelland, and my two children. We were in our Dodge car.

Well, we just kept driving and kept driving, and it had gotten dark. It was so dismal to us that I started off on one of those roads. I realized I had gotten off on the wrong road, and I started to back up, and one of the back wheels spun around and cut the valve stem off the tire. So we had to get under the car and fix it.

That threw us rather late getting into town [Crane]; it was midnight. We came down the main street. My husband was sleeping in a tent behind Dr. Cook's office. That was a doctor who had come there from Levelland. There was no way for me to find where my husband's tent was; there were just tents up and down the street. So we went on.

There was a garage on the corner, and right next to the garage was a little grocery store that had the post office in it, but of course it was closed. Then you skipped two or three lots, and there was a small place that sold newspapers and magazines and that sort of thing. We decided to go down to this newspaper place and ask them where Dr. Cook's office was, because then we could find Roy's tent behind it. We found out nothing—they didn't know where it was—so we got back in the car and backed up.

We could hear somebody down about a block west of where we were. So we just drove down there, and as we drove down, we just drove into a nest of red-light folks that the Rangers were herding down to the square for the night. The Rangers were shooting in the air. They told us to turn around and go back. They didn't have to tell us but once. We got out of that place in a hurry. Then we found somebody that knew Dr. Cook. We went over there and waked them up.

In 1931, **Mary Rogers**' husband had to move to East Texas to keep his job:

He came back home and said, "Well, you better do your washing and get things fixed up: we're going to East Texas." He didn't know whether he wanted to go down there or not. My brother-in-law said, "Let's go down there and look it over. I've been wanting to go down there and see what they have in East Texas. Let's just go down there."

So they went. He called me in a day or two at the little store down at the corner and told me that he was going to stay down there. He said, "I don't know of anything else, another place, with a job and a paycheck coming in." And you know, that was about the first time I ever learned that money was scarce. Not that we were rich or anything.

It was out from Henderson, Texas, at Joinerville. We had a V-8 Ford. And you know they're not very big. So we all got in that car, and we went to East Texas, just to take the children back over there and see the place. There were people living in tents with children. There were a lot of them that had these great big old cardboard boxes draped around trees, living under the trees. And any- and everywhere in the world they could live, they lived. Some were just living in their cars, and a truck if they had a truck.

And I tell you, that was bad. Just no place to stay whatsoever.

We didn't stay that time. We went back over to Mama's. Stayed over there three weeks with them. He called me one night and said, "Well, I've found a little house. It just has two rooms."

I said, "I'll be right there."

He said, "Now, it has a bed and a little dresser and a little stove and an old table and two chairs."

I said, "OK."

He said, "Now, there's not anything else furnished."

I said, "Oh, well, we'll get something to fix us up with."

So I turned around when I got through and told Mama. She went in the kitchen, and she packed up the dishes and a pile of sheets, and some things for the bed. It had a mattress and two pillows, and that was all the bedding in the house.

So we went down there. We carried a lot of things from Mama's over there. Oh, we just got by fine and dandy. Only it was hard.

Vera Lacefield's husband worked in drilling during the 1930s:

I was born and raised in East Texas. I thought when we first got married, "Oh, how rich we're going to be!" It didn't take me long to find out. Well, I was working for fifty cents a day in a grocery store. And I think he made ten or twelve dollars a day as a roughneck. And I thought, "Oh, how much money we're going to have!"

It's just like you have your own little group. We all moved. One would move, well, we'd all move. First time we were out in Hobbs, we didn't stay, I guess not quite a year. John's always—his famous words—mouthing off. I said, "Well, that's one time you mouthed off and got fired." And we went back to East Texas.

Workers, equipment, and teams on
the move in the Burkburnett field in 1918.

We had everything we owned and everything my sister and her husband owned, and a baby; we all moved back to East Texas in this one car. And it was in the wintertime. But my sister and her husband came. He quit too 'cause they didn't have no car or anything, and they didn't want to be left out there.

So we went back to East Texas. But we didn't stay too long that time. I think we moved to Longview. Made a couple of wells.

I've said, "Every town in Texas, Louisiana, Mississippi, where there's an oil well, we went."

Not everyone who had a taste of oil field life stayed with it; for some the uncertain employment and hard living conditions were reason to find some other work. But many workers made the transitions from farm to industry and from the country to the city through the oil field. *John Swendig* offered an explanation:

I think part of the reason that they moved around or were willing to move around is that it was kind of a new found freedom. I mean, they'd been tied to those poor dirt farms all their lives. And they got into a position where they could just up and go. And they did, you know. If they didn't like where they were working or who they were working for, they just packed up and went someplace else.

My father was a farm boy from Imperial, Oklahoma. He did not like farming. But there were so many people there. Those farms were so poor back in those days. I'm sure they didn't have too bad a life. They ate well and everything. But they never had any money. Then when they went into the oil fields, they had money. They didn't have anything else, but they always had money.

Nearly one thousand people got a
free train ride and a free lunch when
the Big Lake Oil Company took
them to see Santa Rita No. 1 gush in
1923.

2: Boomtowns

People who lived in them remember boomtowns as crowded, noisy, and often rowdy places. Teamsters, tank builders, and other off-duty workers drank, gambled, quarreled, and often fought, over the bars, under tables, and in muddy streets. Serious lawlessness, however, was usually confined to the professional criminal element—hijackers, bootleggers, and gamblers. Local lawmen broke up fights and jailed drunken offenders; judges collected small fortunes in fines. But the punishment of more serious offenses was commonly left to the Texas Rangers. The legendary men with the star commonly raided gambling dens, brothels, and speakeasies and rounded up their operators. On some occasions these arrests were so numerous that prisoners were chained to trees and hitching posts because the jails would not hold them all. Most often, the Rangers left town as quickly as they had come, and life returned to its well-worn channels. Rowdiness was still good for business, and it was tolerated, along with most forms of "victimless" crime.

The panorama of an oil boomtown was instantly recognizable: rude, temporary buildings lined narrow, muddy, traffic-clogged streets, on which restaurants, bars, drugstores, and men's clothiers clustered. In the background—and sometimes mingled with the commercial buildings and shotgun houses—tall wooden or steel derricks rose above the make-do sprawl and explained its existence. Old-timers agree that if you saw one boomtown, you saw them all, a claim supported by scores of published photos. Kilgore was much like Burkburnett, which was much like Borger and Wink.

Mrs. *John Berry*: The oil rigs were just like bristles in a hairbrush in Burkburnett in 1918. Just as thick as they could be. There was no spacing whatsoever. It looked like the East Texas field did later. The rigs were all built of lumber, and they all had the big bull wheels. You really have no idea of how dense they were. That's all you saw, just rigs. You looked out the door and it was just like cactus. I've never seen anything else like that.

W. C. *"Bing" Moddox*: Borger was like most all boomtowns. It was made up of corrugated sheet-iron buildings, tents, one-by-twelve lumber shacks, people living in their automobiles and trailers. I'm talking about two-wheel and four-wheel open wagon-bed trailers. And some of them even digging holes back under the caprock and living in caves and half-dugouts.

A typical boomtown tangle of
cars, trucks, horses, and men:
Burkburnett circa 1918.

W. Horace Hickox: What we saw was the real boomtown of Borger. There were cars: you never saw such a traffic jam. There were so many cars! Of course, most of them were Model T's, and some big trucks going down the street. All the streets were dirt streets and had chuckholes. A loaded mail truck actually turned over on Main Street in a chuckhole. Now, that's when I first saw Borger.

Roads between boomtowns and in them were laid out quickly, and they were rarely paved. Heavy use by cars and trucks reduced them to dusty ruts and holes during the dry seasons. When the rains came, streets and roads became lakes of mud.

Carl Angstadt: The roads in Eastland County were bad, almost impassable in 1919. Many a day I've taken all day to get from Eastland to Breckenridge and back. It would take almost all day to drive to Abilene, about half a day to get from Eastland to Ranger. And the mud! I remember a lot of different instances driving along the road and we'd come a to a big mud hole. Couldn't get through. There'd be a farmer sitting out to the side of the mud hole with a team of horses. You'd ask him if he could pull you through. He'd say, "Yes, cost you about five dollars."

Funny thing, after the rain stopped and the mud dried, these farmers would keep filling up these mud holes with water so they'd keep on making their five dollars. I remember a very good friend of mine, contractor, Mr. Lew Teatra, was coming from Breckenridge one day and the roads were all dry but this one particular place. The farmer was sitting out at the side of the road with his team. They wouldn't let you go around the mud hole into their farm; they wouldn't allow you on there. Mr. Teatra drove up to this mud hole and saw that he couldn't get through; asked the farmer if he'd pull him through, and he said, "Yes."

"What'll you charge me?"

"I'll charge you fifteen dollars."

"I've been getting a pull through these mud holes for five dollars."

Axel-deep in mud on an East Texas
road in 1932.

"I'm charging fifteen dollars."

"Well, all right. I guess I'll have to pay it. I've got to get to Eastland. I don't have any money. Would you take a check?"

"Yes. I'll take your check."

And he said he wrote him a check and said: "He pulled me through the mud hole, and I've wondered if he ever found a bank that could read the signature that I put on that check."

Frank Hamilton: I saw them hauling machinery through Burkburnett. They pulled them by oxen and had to float them on floats 'cause everything was so deep in mud. No mule or nothing could stand it. And I know lots of times at Burkburnett they'd carry people across the street. Women would be down there to buy something at that little old store, you know. Well, they had to pick them up and just carry them across there. Wade in mud about up to their knees. If you put any boards across there, why, these big teams come along, oxen and everything.

Wouldn't be nothing there in three minutes after you built a walk there.

Safe drinking water was scarce in boomtowns because utilities were usually built long after town lots were sold and settled.

Dr. D. W. Davis: Our water in Beaumont was cistern water in 1902. Everybody had a cistern and we had to put a strainer on to strain the wiggletails off, and we couldn't always do that. We had a great deal of dysentery, a great deal of typhoid fever, as well as other diseases. Malaria was very prominent. Water was worth more than oil. You could buy water here, but their containers would get contaminated. It was hard to keep them sterile.

Allie V. Scott: Drinking water was one of the biggest problems in McCamey in the 1920s. Even if you boarded, you had to buy your own water. It was one dollar a barrel. They didn't have running water. It was hauled from Alpine by the railroad and unloaded here. There were all kinds of water trucks. They filled up your tank regularly, maybe once a week.

W. Horace Hickox: We didn't even have a water system in Borger. We had thirty-gallon galvanized water cans that we would get full of water. The water man, he'd possibly haul seven hundred gallons of water at a time on his wagon. When he'd come by, he'd stop and we would go out and fill our thirty-gallon water tank. That was our drinking water, our cooking water, our bathing water, if we bathed at home. Most of the men at that time would go downtown to the barbershop.

Sanitation was poor to nonexistent in boomtowns. Few of them had sewer systems before the 1930s; thereafter these facilities were used beyond their capacities during boom times.

W. Horace Hickox: We didn't have flush toilets in Borger. Every house had a privy or a toilet out in the alley. Even the merchants downtown, your department stores, your grocery stores, everybody had a hole, an outdoors privy with a concrete pit under it. That was your toilet system.

As a result, we had rats by the hundreds of thousands. I'm sure, 'cause I saw a great many of them myself. One experience I might relate: I was downtown about eleven o'clock, and I was going down the main street. Not too many people on the sidewalk at that time. And you'd see those rats! They would dart out from between those buildings and run down the sidewalk until they would find an opening between two buildings. Then they'd dart off the sidewalk.

The rats. Golly, there were so many of them that at one time, that summer of 1926, the Rig Theater offered a bounty on rat tails. For ten or twelve rat tails, you could get admission into the theater. Of course, the kids were all trying to trap rats and catch rats to get admission into the theater.

Boomtowns were muddy, unsanitary, and, above all, busy.

A boiler, a ditcher, and a Model T—
all victims of Ranger's famous
muddy streets.

Navigating the main street of
Desdemona after a heavy rain in 1919.

Allie V. Scott: In McCamey, they worked twenty-four hours a day. Everything stayed open twenty-four hours, the eating places and all, because the men worked night shifts and day shifts. I've seen my brother-in-law stay up twenty-four hours at the lumberyard. Businessmen usually had their living quarters at their place of business because of that. They worked Sunday. It was no different from any other day because the oil field operated twenty-four hours a day, seven days a week. You were lucky if you ever had one of the men go to church with you. They had to work. Or they'd be sleeping, if they worked the night before.

Stores, restaurants, and bars were all crowded in boomtowns, but the busiest place in the town was the local post office, as in Borger in 1926.

Gus Keith: It was just flooding. It was one truckload right after another from Amarillo. There were four of us working two windows side by each. General delivery window. And people were lined up for blocks, trying to get their mail. And when every mail would come in, we would have to shut down the windows and then case all this mail into the alphabetical letters and all that, you know. And then whenever we'd get it all cased, we'd open the windows again, and we'd start shuffling mail. And everybody that come up there, they had to stand in line so long, until they would ask their friends or their neighbors or their buddies or whoever. Each person would ask for about five or six different people's mail. And it was a job there, just standing shuffling mail and handing it out all day.

And then we had improvised tables out in the big middle of the workroom there, and when these trucks would come in, they would empty the parcel post or the pouches and distribute it. Mr. Brain, the postmaster, mentioned the fact that they had problems of people breaking into the post office, and the inspectors had been over several times. Mr. Brain suggested, "Why don't you fellows sleep in the post office?" Well, all right. So we got four pulleys, and we put those pulleys up in the ceiling, and we got a pair of springs, and a mattress, a blanket, and a pillow, and we fixed our bed. Then in the daytime we would raise this bed up to the ceiling, and then in the nighttime, when we got through working, why we would lower the bed. And we had shotguns and pistols.

Several decades later, the square around the Scurry County Courthouse teemed with off-duty workers and customers. *Ben Wilson* recalled the hectic days of the Snyder boom:

My auto supply business went through the roof in 1950. Even with the extra man I put on, the line of customers stretched outside my door and on to the sidewalk. I stayed open ten hours a day, six days a week, but that wasn't near enough. You'd be called out regular at night and on Sundays, too. It got to where we didn't have any regular hours.

We had to carry big supplies of long industrial fan belts, bearings, and spark plugs. My inventory was worth $125,000, more than it was in 1978. We were often on back order—of bearings from Dallas, belts from Lubbock, and bolts from Midland. The good part was that delivery was faster then than it is now, or I would've been out of the business.

The south side of Desdemona in 1919: a booming population, a lack of adequate public services, and a random mixture of wells and houses led to sanitation crises.

Desdemona From the South Side

Boom times on Hendrick Boulevard
in Wink, 1928.

Another boomtown crowd: lease
hounds jam an entrance to
Longview's First National Bank in
the early 1930s.

Preboom merchants expanded their stores to accommodate crowds of new customers, and transient storekeepers and peddlers followed the action from one part of the region to another. ***Bill Briggs***, an investigator for Dun and Bradstreet, tracked the boomers across the Southwest during the 1950s:

> There were people who came to boomtowns for short-term gains, in the clothing business particularly; the shoe business was another. There were a lot of filling stations and all kinds of used car dealers.

> Most of these small businessmen were transients. They never even threw up a building. They might use a small mobile home as an office, or they might haul in a small frame shack. All they had to do was get an area graded off level and pack the ground firmly enough so they could put some automobiles on it and they wouldn't sink in the sand up to their hubcaps. None of them stayed long; they'd move on to the next boom and keep following the crests of booms. They'd stay until the drilling activity ebbed and move on to the next play and stay there till that activity ebbed.

> ***Tony Wilburn***: When I came to Wink, I worked in the Sanitary Barbershop. It was anything but! In those days they didn't have the sophisticated equipment they had later. They just used any kind of old junk that they could drag from one oil field to another. The building the shop was in was sheet iron outside; didn't even have any lining on it. It was a store building with a sheet-iron roof. Shiplap inside, Sheetrock, I guess. That's what most of them was built out of. There were seven chairs in the shop. Them boomtown barbershops, they had lots of chairs.

> I went to Snyder and worked during the boom for about three or four months. I don't know whether the shop even had a name or not. I couldn't get in a good shop; they were all sewed up. You'd have had to shoot one of those barbers to get a job. Some fellow came out from Dallas and put in a little two-chair shop out on the Sweetwater highway; I worked for him for a

while. Then I ran the shop while he bootlegged liquor. He decided he'd make more money bootlegging. He was a heck of a swell fellow, but he was trying to make a fast buck. He wasn't an underworld character. He didn't last long at bootlegging. They run him out of town.

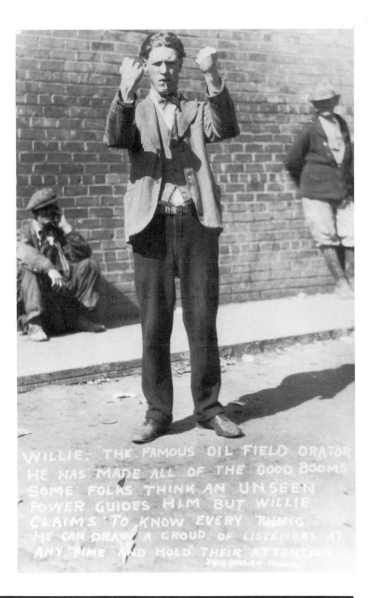

Every oil town had colorful characters like Oil Field Willie, photographed in East Texas during the early 1930s.

Petty criminals, as well as businessmen, traveled from boomtown to boomtown. How much the law let them get away with varied from place to place. During Prohibition, most lawbreaking involved alcohol.

W. A. Allman: In the days of bootlegging, fifty percent of the law were in on the deal. They were paid off. I served one time ten days on a murder trial as a juror. They let us sleep in the sheriff's quarters, which was up in the court-house, and behind his dresser he had a five-gallon jug of the rottenest old corn liquor you ever seen. We didn't drink it 'cause it was rotten. We pulled the cork and smelled it. It was terrible.

Sometimes the Rangers would come in and arrest a few. Sometimes they wouldn't arrest any-one. One time I was working on a wildcat well west of McCamey, and my wife came out in the car with the two children to pick me up at mid-night. We were coming on the highway west of McCamey where the bridge crosses the Pecos River at an angle and you have to slow down to get across the bridge. On the east side was a booze joint. So when we came across the bridge, I looked up on my side and here was a big old .30-30 looking me right straight in the eye. I slammed on my brakes, and it was the Rangers. They were looking for a bootlegger, loaded down with booze. When they looked in the car and saw the two children asleep back there, they sure apologized.

Of course, they were doing their duty. I could have been the bootlegger. If I'd went on fifteen or twenty feet I wonder if I'd been here now, because he had the gun on me. I was looking right down that big old Winchester. I've had fellows say later, "Didn't that make you mad?" I told them, "No, because that's the law's business."

The violence we had in the oil fields was a difference in men's opinions. It was "If I don't like you, I'll whip you, and if you can whip me, fine, we'll shake hands and be friends."

Not a man to argue with: the legendary Texas Ranger Manuel T. "Lone Wolf" Gonzaullas, a celebrity of the East Texas boom.

Hood May: I remember one time I was walking out of Wink there; I don't know, but I guess there was something wrong with my car and I was walking. Well, when I got to the bunkhouse, why, some of them was still up there at the bunkhouse, and we heard a shot down the road there. We run down there, and there was a roustabout got shot, the hijacker had shot him. Well, I'd just been along there. I had seventy, eighty dollars on me. He didn't get me; he got this other guy by the name of Allen, shot him in the shoulder. When he got able, the Gulf gave him a watching job out there at the tank farm, so he could get along. Because that old boy was like me: most of us didn't have much money, you know. So they give him a watching job.

But there at Wink we didn't worry much about hijackers on the rigs. Not out there. Breckenridge was bad about that. They'd go out there; there was chugholes then and you couldn't drive fast. They'd hijack a man while he was on the highway. You had to drive slow; they'd edge up on the car. A lot of them got to thinking it was the law that was doing it. But I never did know.

They had a lot of fistfights, stuff like that. And when I was down there in Ranger, one guy broke a man's neck by hitting him with a fist, actually broke his neck. He was just that good a puncher, you know. That happened right there along the sidewalk right there in Ranger in 1920. They was tough guys like that. Of course, they had to be a-drinking before they was really tough, but they'd usually been drinking.

I never felt I was in any danger, 'cause I sort of stayed away from that bunch. Seemed like I'd pick 'em out. They was tough and I just stayed away from 'em. I'll tell you another thing: if you knew what was going on, you'd better not say anything about it, 'cause you wouldn't live long. And that's the reason I just kept my mouth shut. I knew a lot, but I didn't tell it. 'Cause I wanted to live awhile. And boy, they would, too, they'd get you if you popped off on what they did, bank robbing and anything. So I just shut up about it.

Midland's new Petroleum Building overshadowed the old court house and jail in 1929.

3: Rooms for Rent

Whether they were boomers, drifters, or mobile workers, those who moved with the oil activity had to find a place to sleep wherever they stopped. Before the days of the mobile home, they could not be choosy. Small communities flooded by waves of boomers had few hotel rooms, apartments, or empty houses. Accommodations had to be improvised: virtually any shelter—a spare room, a shed, a loft, a garage, a chicken coop, or a palmetto-covered hut—was rentable at a high price during an oil boom.

After the lucky few snapped up the available rooms and lodgings, later arrivals made do from day to day. They slept in wagons, trucks, or cars, or simply camped outdoors as weather permitted. Unconcerned by the lack of privacy, boomers arriving in Beaumont and Ranger rented chairs in hotel lobbies and barbershops, as well as space on billiard tables, for a night's sleep. Most newcomers eventually found more conventional shelter. Single men frequently lived in barrackslike cot houses, in company-owned bunkhouses, and, more rarely, in hotels. Workers who traveled with wives and children, however, faced the more difficult task of finding something suitable for a family. What they found was often cramped, dirty, and uncomfortable.

The housing best suited to mobile workers and their families was the trailer, now called a mobile home. In the 1920s, homemade wooden boxes on wheels began to appear in the oil patch. Larger and better-designed trailers were everywhere after the late thirties. Following World War II, workers who could afford to buy trailers found them a cleaner and more comfortable alternative to squalid rented rooms.

When growing numbers of oil field workers chose to live in trailer housing, ragtowns and company cottages began to disappear. Even so, trailers have not changed one enduring reality of oil field life: in a booming oil town, a room for the night is still hard to find.

Early C. Deane described roughing it at Humble:

When Humble first hit in 1905, why, they swarmed in there, and [there was] no place to stay. There was no place to sleep much. A fellow would come there looking for work a little ahead of time, and they'd sleep around—it was in the fall—they'd sleep around the gas flare, where the oil went through the separator and into the pit. They'd sleep around it to keep warm at night. Rangers came in there and arrested a lot of them as vagrants. Workmen waiting for work. They weren't vagrants.

In some circumstances, newcomers chose to camp out, like **Bill Briggs** in the early 1950s:

> McCamey had one hotel and a couple of motels that I wouldn't stay in. I would sleep in my car. I would make an arrangement for two dollars a day to use the rest room in a filling station to wash, so that you didn't smell too badly, and shave, and put some shave lotion on, and lots of deodorant. And I'd sleep in the car, eat where I ate, and go on my way.

During the great Snyder boom of the late 1940s, workers camped out and improvised shelter, as **Hugh Boren** recalled:

> There were no motels, hotels, or anything else that could accommodate all the people coming in with oil. At night you could drive out any highway for a distance of ten miles and find people parked on the side of the highway, sleeping in their cars. Some would sleep under the bridges at night. They'd sleep in doorways and back in the alleys. People had no place to stay. They would drive down to Sweetwater or Colorado City or Post or somewhere to eat. A lot of them spent the nights in these towns. It was a very common practice to buy a ticket to the movie and sleep all afternoon.

Those who found beds in cot houses or hotels did not live luxuriously. A cot house commonly had one long undivided room with rows of cots headed along each outside wall. One long aisle ran down the center of the room, from end to end, broken only by a kerosene stove for heat. Cot houses often had running water for washing up, but the toilet was usually "out back." Hotels offered private rooms, though the rooms occasionally lacked doors. Some hotel owners moved hotel buildings from town to town, relocating where the activity was hottest. These mobile hotels changed names as they moved—like some of the promoters and boomers who lived in them.

Palmetto huts provided primitive shelter for workers near the Gulf Coast early in the twentieth century.

Frank Hamilton described one of the first oil field hotels in Texas:

> That Crosby House, it was the largest hotel in Beaumont in 1902. And people were just sitting in chairs and sleeping on top of a desk or anything they could get on the floor. They had these little office spaces that they could stretch out in them, why, they would be sleeping there. And sometimes men would pay as high as a dollar to sleep in a chair. Maybe he'd get ahold of another chair and pull it up and put his feet on it. That would cost him another dollar.

As **Earl Snider** remembered them, the Borger hotels of the late 1920s were primitive:

> They was just sheet-iron buildings that was cardboard, with a long hall. Then it was divided off into rooms. You could sleep in these rooms at night, and when one of these Panhandle dust storms come along, the next morning, why, you got up, shook it out of your hair and kicked it out of your shoes, went ahead and made the best of it. You got your clothes washed wherever you could.

Clell Reed stayed in an equally rugged hotel lodging in Jal, New Mexico, during the mid-thirties:

> We went into the Woolworth Hotel, which was a typical oil field hotel. It was just one long hall with the little rooms on each side. And this old man that ran the hotel was a long, tall, ill man, angry at himself and everybody else. We came in there at five o'clock one morning after working forty-four hours. We rang the bell but no one answered. We laid down to the stove. Wasn't long till he was kicking both of us and wanting to know what we were doing there. Said, "You bums get out of here." My partner immediately grabbed him by the collar and let him know what was taking place; I had to be the mediator. He did finally reconcile himself to where he would rent us a room.

> It was about twelve by twelve. And it had a bed and no running water. You had what they might call a slop jar to answer the call of nature in. It was a bowl and pitcher situation. You poured your water out of the pitcher and washed your face, and that was it.

Supply follows demand as one-room bunkhouses are hauled over the sand dunes to booming Wink in 1928.

Primitive but profitable: a cot house
in Hobbs, New Mexico, in March 1940.

That was a typical oil field hotel. Many, many of them didn't have any outside siding. They just had Sheetrock nailed on the inside.

The great demand for shelter made it possible for newcomers to set up hotels with little capital and even less experience. **H. A. Hedberg** went into the hotel business in the 1920s:

The lease ran out on my business in El Paso, so I decided to give up the business and went on up to Big Lake. I spent the night in what was called the Big Lake Hotel. I noticed that the hotel was cold, and it was not very clean. I had a friend in El Paso who had been in the hotel business and knew the business. So I decided to go back to El Paso to see if I could interest Paul Doran to come with me to Big Lake and look the situation over.

We didn't want to build a hotel; we wanted somebody to erect one, and we could lease it. Nobody was interested in doing that. There was another hotel in Fort Stockton called the Rooney Hotel. We contacted Phil Rooney to see if he would move that to Big Lake, and we would lease it from him. Phil said he didn't have any money to move it, but he would like to entertain the proposition. So we contacted Frank Pickrell, and Frank told Rooney that he would furnish the money to move the hotel to Big Lake. So we leased the hotel and bought the furniture to furnish the hotel.

It was torn down in Fort Stockton, trucked down to Big Lake, and reerected. When it was completed we had fifty rooms, two stories, and a fairly good-sized lobby. It was constructed entirely of wood. We had Simmons steel furniture in the hotel, which we got out of a hotel at Ardmore. It was carpeted throughout. For that part of the country at that time it was a nice hotel. We had an average of eighty to ninety people in it every night.

If the hotel was filled up and an oil scout would say, "Is So-and-so here?" and I'd say, yes, he'd say, "All right, just put me in with him." There was never a closed or locked door in that hotel. It was open day and night. Nobody had keys to

Worn wallboard and a simple tin
roof: an oil field hotel in Hobbs,
March 1940.

the rooms. There was never any thievery. We had a small safe. We had a few slot machines in there too, for the amusement of the boys.

Business boomed so that we erected another hotel, called the Texas Hotel. As the oil business gathered momentum, more and more people came in, and there was more and more business. We were filled continuously. At the end of three years, the boom had passed. The Texas Hotel was torn down and moved to Pyote by a rig builder. From Pyote the hotel was moved to Hobbs. It burned up there, and that was the end of the hotel.

For married couples or families with children, hotel living was not as economical or convenient as apartment living. Ideally, a family hoped to rent a place that would at least allow for some individual privacy and permit cooking simple meals. Most housing fell considerably short of the ideal.

Mrs. I. L. Edwards kept house during the first East Texas oil boom:

Longview was not a big town in 1931. We had a two-room apartment in a big house where there were about four or five other families or couples. It wasn't convenient at all, but that was all we could do for housing. We had two little rooms. We had gas for cooking and heating. The apartment had electricity and a sink with running water in the kitchen, but we had to share one bathroom with all those people. So it was not very convenient, of course.

My husband worked at night a lot and tried to sleep in the daytime. With a child running around, it was not very easy, but we managed. After, I guess, about a year, we built a little house. It was not very much, of course. Everything was very simple in those days. But we did have a place that we could live and be to ourselves. We didn't suffer.

Better-than-average oil field rental
housing, near Snyder in the early 1950s.

Housing continued to be scarce in East Texas after World War II, as **Mrs. E. W. Purdy** described:

> We went to East Texas as two kids out of college. And we didn't have a place to live. Bill went to the rig, and I looked and looked. I finally found a room in a big old East Texas house. The little old lady that still lived in there alone, a widow, had rented out a room on this end, and three more. There was one common bathroom that we all shared. All of us. And our room had a bed and a dresser. There had been a closet, but they'd set a little apartment-sized stove in there. A little bitty cabinet room. No refrigerator. We kept our food in an icebox, just like we'd done in college. We lived there for over a month before I could find an apartment. I went to work for the chamber of commerce on a day basis, just to get out of that little room.

Vera Lacefield faced keeping house in many different oil towns during the thirties and forties:

> We had a one-wheel trailer, and it was built up, covered. And I had those footlocker trunks, and each day everything went in them. It wouldn't take me long 'cause I knew where everything went. It would all stack. All we had was suitcases and boxes.
>
> We'd get a furnished apartment. Some of them would be two or three rooms. Most generally we had private baths. We might have two chairs and a bed. If you were lucky, you had a chest of drawers. And you had a stove. The first time we went out we had an icebox; we didn't have a refrigerator. That's about what you had.
>
> They were terrible places, some of them. Especially in Louisiana. Had so many roaches and ants. I've moved into a lot of dirty rooms and a lot that had roaches and a lot that had mice. Just anything in them! You'd just be there long enough to get it cleaned up, and then you'd go on to another town. Maybe it wouldn't be too far, but you'd just get a place cleaned up and move on. Sometimes it was two or three months. Some of them would last longer than that if there were deep oil wells. We would know from

the time they'd start tearing down the rigs, which would be three or four days.

One time I refused to move into a house. Oh, it was dirty! So we didn't stay in that place but about three weeks. But we did stay, so I decided I was going to have to clean it up. Then my husband didn't have the nerve to come and tell me that we was going to move again, but I knew. When he finally came, I was already packed!

The best way to be sure of having a decent place to live was to buy a trailer. *Martha Lyle* kept house in a trailer during the 1950s:

We started out in a trailer that was twenty-seven feet long and eight feet wide. The table came down from the wall, but it was only when you ate that the table was down. The cabinet was behind, where the dishes were. Had a little stove with two burners on it with a little oven under it. The three older children bunked on the couch in the front of the trailer house and I had a cedar chest that the baby slept on. Then we had one bed in the back. Lamesa was the first place we lived in our trailer. It was the only way we could all be together, to have a place to live as a family. You could not rent a place with four or five children at that time. A lot of trailer camps didn't even want children.

That first trailer was an experience for me. I made the best of it. When it was time for Daddy to come home, I tried to have everything as peaceful as possible, because during the day it was really chaotic. But we were very happy the way we were.

The second trailer we had was thirty-six feet long and had two bedrooms. The last trailer we had had three bedrooms and a bath and all that. We had to work till we got the big one; we had to work for it.

Anne Swendig recalled trailer living at the beginning of the 1960s, far more comfortable than that of earlier decades:

We had a great big forty-two-foot trailer, which we lived in for seven years, and hauled it around when we went to work in the oil field. Our

Trailers, either homemade or manufactured, were the alternative to tents and cot houses in 1940 Hobbs.

Bus apartments for rent at Snyder's
Park Avenue Courts in 1949.

trailer had a kitchen, bath, and a bedroom on each end. Really it was very adequate when you live in this part of the country. We gave it up when we went to North Dakota. I was not going to be cramped up in that with three kids at forty below zero!

It really works out quite well for people who are quite mobile. And for us it worked out very well, because Amerada had a training program they started in 1960 for the new engineers. Instead of having them just do engineering, they had them work in all different areas. They had to do so many months of roustabouting and roughnecking and working in the office and doing geology and well completions and all these things. So they had to switch from one to another. At times I wondered if it was an advantage or a disadvantage for us to have a trailer, because I think when they decided which people would move, I think they said, "Hey, Swendig's got a trailer; they can go anyplace." And I think we did move more often than most of them did. That first year, we moved!

We had a horrible ice storm in Seminole, and I remember being just terrified because when we first went out there, we thought it was going to be for a very short time, like three weeks. If we had had any idea that we were going to be there for six months, we would have done things a little differently. By that time John was getting sick and tired of leveling that trailer house up just so and blocking it, so he did a kind of sloppy job of it. Then we got this horrible ice storm with high winds, and he was out doing well completion and was gone. I'd sit there at night and drink coffee, and the coffee would slop out of the cup because the trailer was rocking back and forth. I can remember lying there in bed that one night trying to map out what I was going to do if it tipped over. The heater was in the middle of the trailer and my bedroom was on one end and the children's on the other end. I thought if it ever tipped over, it would probably catch on fire, and how would I get from my bedroom to the other one? How would I get out of this one window and into the other one?

Another problem that I ran into having a trailer house in the oil field camps was with stoves. We would hitch up to the gas, but it was not regulated as far as pressure goes, like commercial gas is. And stoves would react differently. My oven kind of blew up one day, and I got my hair singed, just because of the difference in pressure.

I enjoyed living in a trailer court; I really did. In fact, I sometimes wish we had never gotten out of them, in a way. Regardless of where you live, if you're in a house there is a certain amount of keeping up with the Joneses. If you have a neighbor who has a lovely yard, you feel kind of guilty, and you've got to get out there and do that. If your neighbor has really nice furniture, it kind of pushes you to get it. When you're in a trailer court, there really isn't too much different that you can do from the next person. There are individual differences, but regardless of the type of job you hold, you're all kind of equal. And you don't have this keeping up with the Joneses. You may have a couple of friends who are professional people and you may have a couple of friends that are laboring people.

4: Three Squares a Day

Food was easier to find than shelter. The rawest boom-town had grocery stores, opened virtually overnight to sell flour, salt, canned peaches, coffee, and other staples. It also had a full complement of cafes, chili parlors, and hamburger stands that stayed open around the clock to feed the changing shifts of hungry workers. Those whose work was far from town had to eat on the job. Drilling contractors often hired black or Mexican cooks who rustled up meals and kept the coffeepot hot. Other employers built boarding-houses, where workers received breakfast, a sack lunch, and supper.

Cooks, waiters, and waitresses followed the booms along with the roughnecks and roustabouts. As soon as they arrived, cooks threw up tents and set up shop, using rough slabs for counters and shipping crates for seating. They cooked on kerosene stoves, and they bought no more food than they could serve in a day, because it could not be stored safely for a longer time. Few short-order cooks, however, were careful with fresh food, and even fewer were particularly clean. Workers ordinarily ordered menu items likely to be relatively fresh, usually eggs or steak. Even then, they often competed with the flies, as **Walter Cline** recalled from his days at Humble, Texas, around 1910:

We were in one of the cafes, the "dirty spoon" down at Humble one day, and there was Mr. Sims, Jim Sharp, and somebody else in the crowd. I think it was Howard Hughes. We had to fan the flies off with one hand and eat with the other. The only safe thing to eat was ham and eggs. You could watch them cut the ham and see them break the eggs. And you had a pretty near even break with the flies from the stove to your plate, you know.

And we were sitting there and eating ham and eggs, and Mr. Sharp said, "Well, there's no use sitting here and not having any fun." He said, "Give me a piece of bread." Gave him a piece of bread and he took some of this salve, looked like axle grease, and it wasn't nearly as good as our present-day oleo, even the cheaper grades. But he took a knife and spread it very carefully over it and cut it into four even-sized pieces. And he put one piece in front of Ed Sims, one in front of Howard Hughes, one in front of himself, and one in front of me. He said, "Now, I'll bet you birds a hundred dollars apiece that there'll be more flies land on my bread before we get through eating than there is on yours, and you can keep your own count."

The name of the winner is lost to posterity.

Major companies often ran boardinghouses for their crews and deducted the cost of meals from their paychecks. Workers for smaller companies could not count on company-run eating places; they commonly bought weekly meal tickets at independently owned boardinghouses. There they ate similar food, but with a more varied crowd of workers, schoolteachers, storekeepers, and, occasionally, prostitutes. By all accounts, everyone ate well.

G. C. "Skeet" McAuley worked in a boardinghouse in Electra in 1913:

> The fellow I worked for owned two boardinghouses at Electra. And I'd stay down there and work for nine dollars a week, with room and board. I made sausage and stuff like that and worked from about five o'clock in the morning to nine at night. You ain't never seen any better eating in your life. They had an area that must have been forty feet long with tables. And they had a big black chef who had cooked in some of the biggest cafes back East in Pittsburgh. And you talk about eating! It would look like that table would sag from the food. And all them oil field hands, you know, they really liked that eating. They had all kinds of vegetables and steaks. At breakfast, they'd have stacks and stacks of eggs and sausage, had everything.

Frank Kelly remembered the routine at Burkburnett in 1921:

> They'd get up at five o'clock and start making breakfast, and they had a huge, great big table—well, it would be larger than from here through this whole house, you know. And it'd be set, just ready for them to come in and sit down. And then they would cook all this stuff; they would cook the bacon and eggs and whatever they put on this table. When the men came in, they'd just go in just like family style. There'd be eggs served, for instance, and maybe two dozen or three dozen eggs on that plate, and you just

The customers look well fed after Sunday dinner at the Morris Boarding House in East Texas during the 1930s.

dipped off what you wanted. They'd just help theirselves. And I can't remember how much the plates were then; seems to me like they were seventy-five cents.

Bessie Leonard worked in the Humble company boardinghouse in McCamey in 1927:

It was just three square meals a day: breakfast, dinner, and supper. We must have been able to feed around fifty to seventy-five men at a time. And I worked in the kitchen, and we'd set up the tables. We set the plates upside down, and then we'd set the cup on top of the plate, and then we turned the saucer upside down over the cup. We wrapped the knife and fork and spoon in a napkin. Believe it or not, Humble sent the laundry to San Angelo to get it done. We used white tablecloths and white napkins in that boardinghouse.

They'd pay about thirty-five cents a meal. Most of them worked for Humble and they had cards, and we just punched the cards. When the card was all gone, they'd go to the company office and buy new cards. We had vegetables come in gallon cans. We just didn't have fresh fruit. You just didn't have lettuce and tomatoes and cantaloupe and beans and peas and radishes. It was all canned. Milk and eggs came in on the train once or twice a week. And the ranchers butchered. We bought straight from them for the boardinghouse. But we had iceboxes, and we couldn't buy too much at a time.

The thousands of cafes in the Texas oil fields did more than serve meals. They extended credit to men and their families, when the first payday was still coming, and they served as clearinghouses and hiring halls for blue-collar workers. **Bobby Weaver** recalled the cafes of Odessa:

The thing I remember was the way the oil field hands went to work. You went to work at the cafe. You didn't go down to the company yard, you went to the cafe.

Dude's was where the roughnecks hung out. Dude's was on Second Street. God, was it big!

The blue plate special—fried chicken, potatoes, red beans, and gravy—at a rig near Kilgore in 1939.

When you went in, you asked Dude if there was any jobs, if you just got in town. Then he'd make you a meal ticket, and he'd get you a job. He got your check; when the check came, Dude got it. You just understood that to pick up your check you went to Dude. Then he'd deduct whatever the meal ticket had accumulated.

The favorite hangout of rig builders was the Carroll Cafe at Tenth and Grant. They hired out there. The roustabouts, general construction work, went to Cookie's. The tank builders hung out at Janie's Cafe. If you wanted to go to work, you stood a better chance at a cafe than by trying to go around to the companies.

Like the roughnecks and roustabouts, the cooks and waitresses commonly did monotonous chores for long hours.

Gus Keith: These Borger cafes were open twenty-four hours a day cooking. Naturally, the people that patronized those places, they would want to eat breakfast, and while they were eating breakfast, the cafe people would fix 'em a box lunch so they could have dinner. Then they would come back and have supper and after supper they were on the town.

Clarence Dunaway, a rig builder, ate at Pinkie's Mecca Cafe in Hobbs, New Mexico, during the late 1930s:

Pinkie was a big, red-faced man. And him and his wife run it. They had a waitress named Pearl. She could wait on more people than half a dozen waitresses. She had about a sixteen-foot shelf back of the counter. And they served what you call short lunch, one sandwich, one piece of pie, and a piece of onion, and a orange or apple, for forty cents. A big lunch was fifty cents; that had two sandwiches. When you ordered bacon and eggs for breakfast, they would have that lunch sitting there before your hot food arrived. That's how fast Pearl was. She never wrote nothing down, didn't use tickets like they do now. She just had a special name for it, and hollered your order out.

Joe and *Theola Starkey* opened an oil field cafe in Monahans, Texas, just after World War II:

We came to Monahans in '47 and bought that little cafe. All we'd ever done was eat in one. We were both bookkeepers. My dad just like to have flipped. He said, "Honey, you're gonna go broke," he said. "You'll go broke just as sure as I'm gone." I said, "Well, if I do, it'll be giving 'em good food. They're gonna get their money's worth." Down there at the cafe I think that God'd taken care of us. We wouldn't have made it if it hadn't been for Him.

But people accepted us. Of course, in any kind of business, you have to sell yourself, just the same as you sell the business. El Paso Natural Gas men, sometimes there's a hundred and fifty of them on a pipeline crew. We could seat twenty-one people. So they sent word to us that they wanted to eat with us. A hundred and fifty. Joe said, "Mom, can we feed 'em?" I said, "Yes." So he sent word back, said, "Send 'em to us twenty-one at a time and we'll feed 'em."

Joe and Theola Starkey served "home cooking" at their Monahans cafe in 1952.

I did the cooking. And of course we had girls that helped.

He never would allow any swearing. This place we bought was a honky-tonk, see; we had a lot of things to live down. They'd come in there and get about two or three sheets in the wind and think they was just real brave and real tough, you know. They'd get out of line; Joe'd tell 'em to get out. If they didn't get out, why, he just picked 'em up and booted 'em out. It didn't take long for that to get around. He made one guy get out one time. He said, "When you can behave yourself, you can come back, but otherwise I better not see you in here." So in about three weeks, "knock, knock" on the back door: it's him. I said, "What you want?"

"Mrs. Joe, would you please let me come in? Is Joe here?"

I said, "No, he isn't."

"Well, let me come in."

I said, "No. I think you owe Joe an apology."

"I can't find him and I want to come in! I'm homesick!"

So when Joe came back, he came back to talk to him. And Joe said, "If you think you can behave yourself, you can come back. I told you that. Son, this is a family cafe. My girls, you're gonna treat 'em with respect, or else you're not coming in here."

And the girls just loved him to death after that.

We usually opened at six, and at eight we shut the doors, 'cause they'd be ding-donging at that front door, we never would get out of there. And if we'd take the notion, we'd shut down and go fishing. They'd say it was the only place in the world that could shut down and come back and have this many customers when they come back.

I cooked whatever I'd take a notion to cook. Chicken-fried steak: you serve a man a chicken-fried steak and some red beans! I fixed chicken-fried steak, roast beef, baked ham, hamburger steaks. And then I had vegetables. Those boys that are out in the field, they need those vegetables. The first time, in 1947, we got seventy-five cents for a meal, and that included the drink and dessert. And we had all we could do. And when we closed it up in 1978, I was getting two-fifty, and having a hard time breaking even at that. And those old boys eat!

We got a lot of them jobs. The companies would telephone down there for hands.

I felt like we done a little bit of good in Monahans. Because, some time back, I was walking down the street, and this lady said, "Aren't you Mrs. Joe?"

And I said, "Yes, Mrs. Joe Starkey."

She says, "Do you remember me?"

I says, "No, honey, I sure don't. You tell me. Maybe I can."

She says, "When we came to town, we were broke, and we didn't have a place to stay, and our baby was sick. Mr. Joe said, 'Well, we haven't done a good deed today; those meals are on us.' And you helped us find a place to stay and called the doctor and stood good for the doctor bill."

Then of course I remembered events. Every bit of it was paid; they paid back every bit.

Many of the owners of the stands and cafes in oil towns are remembered as colorful characters; they were some of the most memorable people in the oil patch. One of them was a pioneer in fast-food sales in Borger, as **W. Horace Hickox** remembered:

Chink-Link was Borger's most popular hamburger stand in 1926. It was run by Pappy Oric and he sold hamburgers for a nickel. Now, they weren't jumbo hamburgers, but he sold them for a nickel. His slogan was "A fast nickel beats a slow dime!" He stayed in business, in good business, until about the time the war started. Meat got hard to get. That's about the time he closed out.

C. O. Puckett: We went into a little cafe in Hobbs at about two o'clock in the morning,

Ready for a break, rig builders stop
for lunch at Humble in 1907.

that's in 1926. It was just a boxcar shack with stools down the counter. I'll never forget how dirty the cook looked. But we were hungry. That guy had hairy arms and a dirty apron. He had a week's growth of beard. Rough-looking. Some guy sittin' down the way, said something about the food put out for him. This cook just put his hands on the counter and put his ugly face across it and said, "By God, you ordered it, and by God, you'll eat it."

George Mitchell: There was an old lady came to Kermit during the 1930s; we called her Ma Nations. She was a corker. She owned a traveling circus in the summer and in the winter she'd come to Kermit to cook. She put the lights she had for her sideshow on one of them little old two-by-four Sheetrock buildings down there on Main Street, in what they called Hell's Half Acre. She called it the Electric Kitchen.

Old Abie was drilling a well out here one time. He told Bill Clinton, the night clerk at the Texas Hotel, "Come on out to the lease with me." They'd got a rig running. On the way, they stopped by the Electric Kitchen, because Bill had never seen the little old pet monkey this woman kept on her shoulder. So Abie went in and took off his hard sailor hat off his bald head. They sat down at the counter and ordered a bowl of chili apiece, but Bill didn't want to eat anything in there. Nobody hardly would eat in there except them circus hands. But Abie ordered a bowl of chili. Then the action started.

About the time the waiter was starting out of the kitchen with the chili, Ma Nations came in with that monkey on her shoulder. At the same time, somebody come in the front door and let an old house cat in. The cat ran behind the counter; the monkey spotted it and bailed off Ma's shoulder. There was a quick free-for-all about the time the waiter got to Abie with the chili. Old Abie raised up to look behind the counter to see what the fracas was; when he did it, the waiter turned to watch it, too, and poured that bowl of chili right over Abie's bald head. He didn't go back in there anymore.

Clean windows and fast food:
Snyder's new drive-in in the early 1950s.

Blackie Becker, widely known as Oil Field Blackie, started cooking near Houston before World War I. By the late 1920s, he had moved to Breckenridge and on to McCamey, where he ran a place with his name over the door.

Cullen Akins: Blackie was quite a character; a man who looked the part of an oil field promoter and had a mouthful of gold teeth. He was a past master at punching meal tickets. They told me that was the only place to eat, but it didn't take long to find another place because I had to pass right by his kitchen every day going to the post office. He had a cook who was a retired Navy man. This cook had a parrot and that parrot was pretty handy about walking over the dishes, so I found another place to eat.

Don Dittman: He fed all the oil field hands all during the boom and the Depression. He closed up sometime in the Depression. You'd go in there and say you wanted to eat on credit, and there was no questions asked. But you didn't beat him but once. He had followed the oil fields all over the country, and you didn't beat him but once.

His place was just a typical old-style cafe. They had family-style meals. They just set it out on the long tables, and you ate just what you wanted. Most of the time they just had meat and potatoes and gravy and beans and stuff like that. Stuff that the workingman required.

Mrs. J. M. Horner: He never turned a man away if he was hungry, and he was kind to children. That little girl of mine met him not so long ago up at Lubbock after she had been working for the Humble Oil Company. He remembered what she always ordered when she went into his cafe, when she was six years old. It was pork chops and fried potatoes. He never forgot that. He was popular and he cooked good meals.

Killing thirst near Kilgore, April 1939.

5: Ragtowns

Before modern mobile homes appeared in the oil field, shacks and tents were the most common temporary housing. Hastily erected shacks, thrown together with whatever materials came to hand, were commonplace from the early Pennsylvania discoveries onward. As an oil discovery brought a rush of newcomers to an area, shack and tent settlements sprang up like mushrooms after a rain; because of the many canvas tent tops, these instant settlements were called ragtowns. The term was especially common after World War I, when Army surplus tents became available.

Most ragtowns were short-lived. They met temporary housing needs, and those who lived in them ordinarily stayed only as long as an oil boom lasted. When drilling slumped, workers moved on, taking their tents with them. A few ragtowns lasted longer than others because they were part of company camps. These poor-boy camps, as workers called them, were built on company land. The residents, blue-collar production workers for the most part, were supplied with water, gas, and electricity by the company, but they put up their own shacks and tents. In time, these workers either moved on or built up enough seniority to claim better and more permanent housing in the company camp. In the meantime, the workers and their families put up with the inconveniences and hardship common to both poor-boy camps and ragtown life.

Tents and shacks kept out most of the rain, but not much else. The hard-driving winds of North Texas, the Panhandle, and West Texas whistled through tent flaps and cracks in walls, carrying soil and sand over every surface and into every corner. In East Texas and along the Gulf Coast, swarms of mosquitoes and flies tormented ragtown residents, spreading disease as they went. Few camps enjoyed the luxury of running water until the 1920s. Bathing was done in a galvanized washtub or under a bucket shower. Sanitary facilities consisted of a two-holer somewhere out back.

Ragtown living was harder on wives and mothers than it was on husbands and children. The workers were gone from the make-do shelters most of the day. The older children went to school and generally escaped the responsibility and worry that accompanied holding a family together under difficult conditions. But for the wives and mothers, there was no escape from the constant problem of trying to transform a tent or a shack into an acceptable home. Housekeeping conditions were primitive, quarters were cramped, and small children were always underfoot.

The greatest aids to coping with ragtown housekeeping were faith that it was only a temporary detour on the road to a better life and knowledge that neighbors shared the same hardships. Sometimes common hopes and adversity even fostered a sense of neighborliness, of belonging. Onetime ragtown residents generally recall making friends, sharing simple fun, and caring for neighbors in times of personal crisis. They remember the primitive settlements with some nostalgia, as places in which human relationships were more important than material possessions. Perhaps these positive and sentimental recollections are natural to the survivors of that experience because they recall the days when they were young and the promise of a better life lay ahead of them.

Tony Wilburn first lived in a ragtown when he came to Pyote in the late 1920s:

> I stayed in Pyote nearly a year, and I slept in a tent house over there. I wasn't alone either. There was 3,500 people lived in Pyote then, and some said that 3,000 lived in tent rag houses. They called those tents rag houses. There weren't any houses.

The way they fixed those things, they would put a wooden floor and wood around, up to about three and a half or four feet all around. Then they put a tent or tarpaulin and fastened [it] down just below the top of this wood. Then they'd screen it in. They had to on account of the flies. They had screens up to the top.

And that was the sweetest sleeping you ever had. At night you'd just roll that tarpaulin up, and that breeze would just come through there, and it was the finest sleeping you ever see. Just a little bit shaky when it was stormy.

Bessie Leonard kept house in a tent in McCamey early in 1926:

> My father-in-law was already there, and he knew we were coming. He had bought a tent from someone who had said, "Oh, we're not staying here! We're going on." They just went on to California or somewhere. So we was fortunate: we had a tent to move right into when we got

here. Otherwise I guess we would have camped out just like a lot of other people did.

We finally boxed our tents up, but when we first lived in them, they weren't boxed. You had all kinds of little visitors that came in. We had mosquito netting. We couldn't have slept a night if we hadn't had it. It was just rolls and rolls of mosquito netting, and you tied it. Our bedposts was set in oil to keep ants from getting on the bed. There was flying ants, there was flying other little bugs, and centipedes was bad, and just a lot of hazardous things.

We had many, many meals with our heads under a sheet. We'd put the food on the table, on boards or whatever, you know. You would fix your meal and cook everything covered to keep the sand out. Then you would eat with your head in under a sheet. We put the table legs in kerosene to keep the ants and other stuff from coming up.

We had pack rats. They carried away just about everything that was loose. But if you'd go out to where they had their den, you'd find it. If you kept all of your garbage and things carried away, they weren't too bad.

At first we didn't even have gas. They had gas wells, they was bringing 'em in, but they hadn't piped it where people could use it yet, you know. You had Dutch ovens that you made your biscuits and cornbread in. At that time I had one that had a rim up around it. I baked my pies, cakes, light bread, everything in it. I used to sit it on hot coals and put hot coals up over the top.

We had to buy water for everything. For two, maybe three years we had to buy water. We'd pay one dollar a barrel for it. If the train didn't come in with the water, it was just terrible. You took your bath in it, but you didn't throw it out. You kept it for your clothes, 'cause your clothes was full of sand. You'd put them on the line, and you'd hope the sand wouldn't start blowing and get them full of sand until you'd get them off the line.

Water was very precious. You could go down to the Pecos River, we'd go down to the river and take baths, but it was awful gyppy. When you got out of it, you felt sticky. But nevertheless, you felt cleaner. You got the grit off you.

For a privy, you put a hole in a board, and that's what you used. Everything went out on the ground. It was so sandy and everything, you didn't put a box in. Everything seemed to disappear. I guess maybe bugs eat it, I don't know. It was amazing. I had been used to modern plumbing. When you went out and looked, that waste was gone.

But really—I was thinking of it today—it's odd. Today we lock the doors, be sure our screens are fastened, and most usually we pull our windows down. Them days your husband went to work in the evening. They worked about twelve hours in them days, didn't have eight-hour days. You'd go to bed with that open tent. You never thought a thing in the world about it. You was never molested in any way whatsoever.

Mr. and Mrs. R. V. Melton had more furniture than most tent dwellers in their tent in Crane in 1927.

R. V. Melton: We had a mansion here. We had a twenty-by-twenty-four-foot Army supply tent with a good floor and good walls. Most of them had little old eight-by-twelve tents, but we had a big one.

Mrs. R. V. Melton: We had a two-burner oil stove, and we had two beds and a baby bed, a Victrola, a sewing machine, a dresser, table, and chairs. At one time I believe we had three beds in that one tent. On several occasions, on this little two-burner oil stove, I prepared chicken, dressing, and all the trimmings, and we had a good feast with company in. We didn't have refrigeration. Most people would buy ice from the

A typical ragtown horizon of tents, shacks, and derricks lies behind mule teamsters in an early Texas oil town.

iceman every day. The iceman would bring the ice, and we would get ten cents' worth. That would make tea for our dinner and supper; and we would wrap it in a quilt, and keep it in a tub. If you had butter and milk, you would put that real close to the ice. We kept the ice in a number-three tub: that same tub was your bathtub. On every bath occasion, you emptied your ice and everything out of the tub, took your bath; then you cleaned your tub out.

Drilling contractors who brought workers to remote wildcat locations put up temporary, ragtown camps for workers. *Vera Lacefield* described living in a company-supplied tent in such a camp in Pecos County, in the mid-1930s:

We stayed down there about a year and a half, and everybody lived in tents. [Ours] was about twelve feet by twelve feet; we had it built up and the tent part on top. We'd roll up the side, that was our ventilation. [In the winter] it got cold; we had to have a coal oil heating stove.

We had an outdoor toilet. They had a bath fixed that was, oh, a barrel with water in it. And we had to haul all our drinking water. We washed outside of the tent on a rub board. The day we'd go get water, had clean water, that day you washed. But I don't know, I guess when you have to do something like that, you can manage.

We had the prettiest cactus bed. We'd roam around, hunt cactus. That's all we had to do. Wasn't nothing for entertainment. A lot of the women played bridge and passed the day away. [The men] didn't have far to go to work, and lots of times we'd take the kids and go down close to the rig, and set and watch them. We had a boardinghouse, and everybody ate there. Some of the women that really wanted to, went over and helped in the kitchen, like serving and things. I never did.

We went to Iraan or Fort Stockton. It was just over the mountain, but we had to go all the way around to get to it, on a little goat trail. Take all

Two sisters pose by their family's boxed-up tent near Andrews in the late 1930s. It was probably necessary to move some of the furniture outside, though the tent was larger than most.

day to go in and back because the roads were so rough. And the sheep wouldn't get out of the way, and you couldn't run over them!

But I guess it could have been worse. Back then, really, we were glad to have a job.

The Southern Crude Oil Purchasing Company had a camp in Wink; many workers who lived there during the late twenties and early thirties spent several years in tents. *Alice Keene* still lives near the location of her tent in 1927:

It was my first experience to ever live in a tent. We got our tents, big tents. Fixed 'em together, and they were walled up. All of us that could boxed up our tents.

Well, you know, the wind would get up in the night, and many, many times we'd go to bed and have the curtains up on the tent: sandstorm would come, and the sand would be under us. One night that big boy says, "Mom, we're going to set the table so we don't have to come down so early in the morning." I said, "Well, you better watch out." Anyway, they set the table. They turned the plates upside down over the knives and forks; they put the preserves and jellies out, the ones that had covers on 'em. Next morning you couldn't tell where the plates were! They came down and said, "Can we come in?" And I said, "You've got a job. I'm not going to touch that table. That was your plan; now you can clean it up."

Oh, it'd sand! We had the awfullest sandstorms. And not being used to it, my hands got so sore, my fingers and all. Every time it'd sand, I'd think I had to clean it out. We didn't have too much furniture or anything, but you wanted to keep clean what you had. Oh, it was terrible in those days, I tell you!

It was years before we got any plumbing inside. We had johnnies. It'd be—maybe three or four families would use it. It would be the two-holers; some of them would be large enough that they'd have three. We got us a piece of oilcloth, and when we'd go, we used our piece of oilcloth over these, because you never knew who was using them, what happened, or anything. You didn't know what you'd get. But that's the way we protected ourself.

We had a problem with flies. But really, I can't remember seeing a roach in my house in those days. You have to watch out for things like that now more than you did in the early days.

I did all the big laundry. We had to do it outside on the rub board. We had a table fixed, we'd lay these oily clothes on there and then soak them with kerosene. And we had a drum had been cut in two. They fixed it so you had gas fixed out there, and then we'd boil [the clothes]. Sometimes, if you didn't watch, you'd catch 'em on fire from the kerosene; it'd be on the water, and the flame would send 'em up. [If a sandstorm blew in], well, we'd put our clothes in the tub, in the rinse water, and take towels and things and cover them over, thinking we'd keep the dirt out. After the sandstorm was over, why, we had them all to wash over again. Sometimes the line would break, and oh, we had a time!

Do you know, though, those were the happiest days!

As a child, **Ruth Godwin** was one of Alice Keene's neighbors in the same camp:

When we first moved to Wink, we lived in two tents: we didn't move to a house till 1931. There was no rent or anything like that. We owned our tents. They were walled up about four feet on each side and had floors and walls. Some tents had pitched roofs and others were pyramidal. The pyramidal tents were the ones that suffered so badly during our windstorms.

We had one tent in the front that had the cook stove and the cabinets of sorts and a double bed. In the back tent there were four bunk beds. We actually had more space than many people had.

We had gas piped in as fuel. We used our gas stove for heating as well as cooking. Most of the stoves were little cook stoves with the gas piped in. Sometimes when something happened where

Boxcar houses were hauled from place to place. This Odessa house was still in use in 1980.

they separated the oil and the gas [in the field], the oil got in the [gas] line, and we had the pleasure of cleaning up the mess from that oil that was burning in the stoves. It was quite messy at times.

There was an ice plant in Wink; that was one of the early things that came in there. We bought ice and used an icebox for a long time. We had water piped right to the tent. We had plenty of water.

At that time in the camps they had community showers and bathrooms at various places in the camps. For every so many tents, they'd have a bathhouse. They had one for the men and one for the women, but they were stationed all along our campsite. You had to take your turn at taking showers out there.

It's unusual. We had come from a better housing situation. We came from New Mexico; my dad had worked for the Santa Fe Railroad. Mother had a great number of lovely things that she'd had all her life. She brought her things along, but we had no place to store them or anything in our tents. This gorgeous cut-glass punch bowl: we finally wound up using it for a fishbowl, I think. A great many of her nice things were misused, that's for sure. But that was true with most people.

It must have been frustrating for people who had not lived like that, though, had not lived in tents. I wonder how my mother stood it. She was a college graduate. And I think she must have wondered how in the world she ever got to Wink.

Ragtown dwellers died as they lived, underneath a canvas roof. **Mrs. R. V. Wilson** recalled the death of a neighbor in Crane:

We had a man that lived a block in front of us, and he died, I believe, of typhoid fever. That was the first time my children ever saw a corpse. All they did was raise the tent side up and laid him out on the boards. We didn't have an undertaker. The nearest one was in Midland.

They just raised the side of the tent. They didn't have screens or anything. My children went down there and saw where they had put his false teeth back into his mouth, and—can you believe it?—they were just glaring. They didn't fit very good after he was dead. That, of course, impressed me more than anything.

They were poor people, and they just built a wooden box. The women put cotton padding in it, and a thin cloth of some kind, like lawn, and lined the casket. The men just put him in the casket and nailed it up. We didn't have an undertaker for him at all.

Then they had the funeral. [There were] just these men to carry the casket, and they took it out to the graveyard and buried it. The men had dug the grave. The family all came out, and the neighbors. They always had a service at the grave.

Listen, I think that the people in Crane, those that were buried in pine boxes, were just as well buried as some today are in these big, expensive caskets; in fact, I don't believe in them.

When tents were not available, oil field workers sometimes constructed shelter from unorthodox building materials. **Mrs. James Williams, Sr.**, remembered:

During the Depression, while we were drilling that well out at McCamey, we lived in a little tin house out there that we made out of these old barrels that the oil companies throw away. We made a house out of those, and lived in it with the three children. We had thirty dollars a month to live on, but we had a cow, and between that thirty dollars and the cow—of course, we had to feed the cow out of the thirty dollars—we had plenty of milk and butter and all the good things you make from cream. Our icebox was a hole in the ground. They put boards around it and left a small opening all the way around the board. Then they stuffed that full of sawdust. We wet that, and we put ice in it, and it kept just as good as an icebox. We kept fresh vegetables and milk and everything in it.

We didn't suffer from it. You had to be satisfied, with my husband; he wouldn't let you be otherwise! I think I could do it again without any trouble.

W. Horace Hickox remembered similar shelter in the Phillips ragtown near Borger:

It was tin shacks, shacks and little tents. And I might mention that lots of those houses were built out of old carbon-black tents, these carbon-black burner houses that they had, the carbon companies had. When the tent would get old and rusty, they would replace it. This old town was junk. They would take and build houses out of it [discarded carbon-black tents], and then they would stucco over it or something; some of them would and some of them wouldn't. There were lots of those houses on Ragtown built out of old carbide cans. They would take those old carbide cans and barrels and cut them open, tack them on the walls. Most of them didn't get that coat of stucco that dressed them up a little bit. Nine out of ten didn't. So that's what you saw when you went into Ragtown and much, much of Borger.

Workers who could afford lumber often built themselves shacks, selling them to other workers when it was time to move on. *John Rust* recalled such housing in Ranger in 1919:

Suppose a rig building crew come in there, and they needed housing for themselves and their families. The first thing they'd do was put up four or five or as many of those little shotgun shacks as they needed and move in. They're built [with the rooms] in a straight line, just one room behind the other, like a long-barreled hog rifle.

They built their houses on rented ground. They'd pay five to ten dollars a month for ground rent, and then they'd live there as long as they needed to live there. They'd move away, someone else would move in and continue that ground rent. That was all the landlord ever seemed to want, was just his ground rent.

On my daddy's land, he had hundreds and hundreds of these shotgun shacks all over, on eighty acres of land. Much later on, when the

A ragtown near Odessa in 1937: at
left is the privy shared by the shack dwellers.

people began to move away, we had a lot of those empty shacks on the place. But I'll tell you this: he, as well as many others there at Ranger, he made lots of money on ground rent for those little shacks.

Mr. and Mrs. *Hoke Tehee* lived in the Corsicana field in 1922.

Hoke Tehee: We got down there at that Corsicana, I think we got in there along about October, November, or something, and it started raining there. It rained all winter long. The mud, some places it got two feet deep. We lived in a tent at that Corsicana, lived in it all winter. Then I built us a little old room, I put together a shack. Cost $150. Just one-by-twelves on the outside. It had two or three windows and a front door. Sold it for $200 when we had to leave, so I made a little profit on it. We didn't move anything. We just packed a suitcase and sold everything we had in the house, got rid of it. We'd just take what we could get in the car.

Mrs. Hoke Tehee: We had one child born in Corsicana. We was happy. We was kids.

Skeet McAuley built his family a home in Wickett in the mid-thirties:

That house cost me about $385 by the time I got done with it. I built it myself. [The rooms] was about ten by twelve, [the ceiling] was eight foot. I framed it, and I put that quarter-inch plywood on the inside. I built cabinets and paneled the doors on it. Did it all with a saw and a four-inch wood chisel and a hammer. That was all the tools I had.

Mrs. Skeet McAuley: He built that house, shotgun house. Of course we had a kitchen, and he made a great long table, just as long as the room almost, and benches on either side. I had a stove and an icebox; we had to go get the ice. The front bedroom just had a bed and a chest of drawers, and then I had a baby bed [there], too. The middle bedroom was where our [nine] children slept. And in between those two bedrooms

The house that Skeet built: Mr. and
Mrs. G. C. McAuley pose in front of
the plywood house he constructed in Wickett.

there was maybe a four-foot closet on either
side. One was the men's and one was the ladies'.
We didn't have many clothes, you know. One
Christmas he was going to get everybody Christmas presents with whatever money he had; he
didn't have very much. And he come back with
balls; got that for every one of them. He hid
them with their clothes. Somebody opened the
closet door, and here go all the balls! And he was
so disappointed—there went the surprise!

Those who came to oil towns only to find that the available housing consisted of slapped-together shacks were
often less enthusiastic about their shelter than the original
builders. **Dr. Homer Johnson** described the housing he
found in Crane in 1937 when he arrived to practice
medicine:

We found a terrible place to live. It was two
little buildings that had been put together in a
T-shape. It was a tin-roofed, wood-frame building. The floors on the house had big cracks; you
could see through it and everything. It was
terrible.

My wife went out and she looked at it—didn't like that very good! She thought, "Well, we'll try it for a little while and see how we'll get by." We left most of our good furniture in Abilene with my wife's mother; we did carry some carpets and put [them] on the floors. We went out there to stay a year, and we stayed about fourteen months.

The wife, one time she was getting ready to come down to the office and stay with me. She was getting dressed there, and a big sandstorm, she saw it coming up. That thing hit. The wind blew so hard underneath this house, picked this thick old carpet up, just wrapped it all around her, scared her to death. Boy, she like to have had a fit! She was ready to go back to Abilene or anywhere else.

It was terrible. I don't know how she put up with it. But we did meet some of the most wonderful people that I ever met in my whole life.

Mrs. Clell Reed remembered similar housing in Wink:

Humble owned the land our houses were on. It was in what they called poor-boys' camp. It was just a shack. We had a bedroom and a living room and a kitchen. When we moved in, we had an outdoor toilet, but they put plumbing in while we lived there. No closet doors: I had a curtain over the closet. We had linoleum on the kitchen floor, and it would blow up so high we would just have to mash it down step by step as we'd go. Oh, nobody knows what people went through and lived with at that time.

We invited [other family members] all out one year for Thanksgiving, and all Clell's family came. We think back on it now and think how sorry they must have all felt for us, because it rained that morning at five o'clock, and the electricity went off, and everything was out. I had my turkey to cook and everything. Oh, I don't know how they stood it, but they did. We have often laughed and laughed about what they were thinking. We were so proud to have what we had and so grateful to have what we had.

Relatives from home, as **Mary Rogers** recalled, had a hard time understanding oil field housing:

> My mama never could understand, never did understand about the oil field people. They spent their money on the families. They'd have a car. And of course radios; well, anything and everything. And didn't have a house. Now, if my mama loved anything—and usually had a good one—it was a house.

Despite their ramshackle shelter, many ragtown dwellers still remember having good times; young and hopeful about the future, they did not let living in a tent or shack keep them from enjoying Saturday night. **Mrs. R. V. Wilson** summed up a common perspective:

> Ragtown was halfway between the Gulf camp and town [Crane in the late 1920s]. That's where we lived first in a shack. After my husband built this fourteen-by-twenty building, he had some wood left over, and he built a trundle bed for the children and our bed. At night we would have to set the two rockers out in the front yard and pull the trundle bed out. Don't think for a minute we didn't have company. We had a lean-to tent on the side of the house, and a sanitary couch. You would catch ahold of it and pull it, and it would come out and make a double bed.
>
> This fourteen-by-twenty room we had, I had a kitchen cabinet called a Hoosier. They were the nicest kitchen cabinets you could find. I wouldn't let anybody move that kitchen cabinet, but I'd let them move all the rest of the furniture out of my house, and we would have a

Laundry day in an Andrews ragtown
of the 1930s: the clothes were boiled
in the sawed-off oil barrel at lower left.

Borger, 1926: there was always
strong demand for this ragtown service.

party. Some of them would pick banjos, ukeleles, and guitars. We would dance and just have the best time. When they got through, we would move all my stuff back into that room.

We just had the best of times. You just can't imagine how much fun we had.

Five rooms with bath and a screened
front porch: Humble Pipe Line's
Kemper Station Camp in 1925.

6: Company Camps

In the scramble for oil field housing, the winners got company houses in company camps. At the beginning of the century, these camps were temporary, meant to last for the duration of work in one field. Early company camps served only to give shelter to employees who worked too far from towns to find housing in more conventional surroundings. By the 1920s, many large oil companies came to provide more than rough shelter. Following the examples of Eastern textile mills and Western mining companies, they offered appealing alternatives to primitive living conditions in order to attract and keep trained, reliable workers. By providing middle-class comforts, the companies headed off complaints from wives that the oil field was no place to raise a family. The provision of decent housing was also seen as a way to build the kind of loyalty to the company that led to higher productivity on the job and discouraged support for the organizing efforts of labor unions. Housing became a necessary and attractive fringe benefit, and company camps became a familiar feature of the American oil field.

Company camps varied in size, but they looked alike. The houses, painted in some uniform color scheme—white with green trim was most common—were laid out in orderly rows along streets maintained by the company. The company ordinarily planted grass, trees, and a flower bed or two in the camp, and when maintenance crews happened to be idle, they were put to work mowing grass and trimming shrubs. Depending on the size of the camp, it might contain divisional offices, warehouses, garages, and repair shops, all handling day-to-day management and operations. Most camps lacked retail stores or shops; to buy a pound of bacon or a pair of shoes, to have a haircut or get a suit cleaned, camp dwellers had to go to the nearest town.

Compared to the tents and shacks of ragtowns, the simple but sturdily built houses of company camps were luxurious. The average house had two or three bedrooms, a living room, a small dining area, a kitchen, and a bathroom. Householders paid only a nominal rent and a small fee for company-supplied utilities. Both were deducted from their paychecks. The company did all repair and maintenance work; sometimes it did yardwork as well, though in most camps householders mowed their own lawns and kept their own gardens.

The company camp was a secure, self-contained world in which a family's identity and status were defined by the job of the breadwinner. Its advantages were numerous, and most residents enjoyed camp life. Some, however, found the requirements of loyalty to the company and conformity with the values and social systems of the camps too confining: for them camp life meant a loss of freedom.

H. P. Slagel and William H. "Bill" Collyns lived in oil camps in the twenties.

H. P. Slagel: The purpose of the camp was so a person could get some rest. If you lived in Colorado City, you know, it was ten miles out there to that well. The roads didn't have to be too bad so it would take you three hours to drive out there and three hours to drive back. Then you worked twelve hours, and you didn't have much time to play around. That's where the camps came in. There have been times when it was all you could do to drive from there into town and get something to eat and get back out there in twelve hours. So there were camps all over the field. They were simple but adequate.

Bill Collyns: Camp life was just a part of the game then. I don't think they could have gotten employees out there if they hadn't had the camps.

I don't think I would have gone out there. I wouldn't have had anyplace to live, and I couldn't have afforded outside housing—well, there just wasn't any to speak of.

McCamey was a comparatively young town at that time [1928], a boomtown I guess. The Humble had established a camp, as did Shell and several other companies. Humble's was one of the largest camps. They built a division office building, and the Humble Pipe Line Company also had built an office. They had their own Humble Pipe Line camp. The [Humble] refinery was there. It was only a short distance from the camps, a half-mile or so; it had its own camp also.

The production camp must have had fifty or sixty houses. The production camp and the pipeline camp were very fine facilities. There were anywhere from five- to seven- or eight-room homes. The superintendent, the division superintendent, and the chief geologist had very nice homes. They were rented to employees very cheaply. Rent was very cheap.

I first lived in the bunkhouse. We had bunk-houses for the single men. They were long-type houses with a porch. They had about three in the production camp, three bunkhouses; they had a lot of single people work there. We lived two to a room and, I think, twenty rooms to a bunkhouse. They were fairly good-sized rooms. They had two beds, and a large table in be-tween, and a washbasin, and that sort of thing. It wasn't spacious quarters at all, but very com-fortable quarters. And in McCamey, in that day, it was about as comfortable as you could get. When we lived in those bunkhouses, the rate was very minimal, maybe a dollar a month or something like that. They'd deduct that.

After five or six years, my mother and my sister came out from San Angelo and joined me, and we rented one of the houses. It was a five-room house. We had gas, but electricity was very cheap at that time in McCamey, and I re-member we had an electric water heater. We had a water system that the company kept up. We had a front room, two bedrooms, a kitchen, a dining room, and a single bath.

Humble set out a lot of trees, and that was about the extent of it. Then the residents of the camp kept up their own lawns. Some had nice gardens, nice flower beds: very attractive. And others weren't inclined to yardwork. (I am not inclined to yardwork!) The road surfaces were paved in the camp. First gravel and then black-top. They kept them in good shape. I remember when they spent considerable time and effort and money on landscaping, particularly grassing the area. Had Bermuda grass. I remember writ-ing a story about the first lawn mower used in McCamey; that was used in the Humble camp. They had some of their own water wells, and that was pretty good water for irrigation.

It was kind of a family deal, actually. Every-body kind of knew what was going on. I think that might have been one reason they finally de-cided to do away with camps: some of them knew too much! But in those days, I don't recall any internal troubles in the camps. Everybody seemed to enjoy themselves. They had friends and neighbors. They had quite a bit of home brew making at about that time. They'd have home brew, have a good time, and parties. The production camp had a tennis court. And once a year we'd have a holiday, Humble Day, they called it. All the employees and their families would have a big picnic and games of various kinds, horseshoe pitching.

So, it was about as nice a facility as you could find.

At Texon, the Big Lake Oil Company ran a camp for its employees that was more like a company town than a company camp. Mr. and Mrs. Joe Koesel lived there during the 1930s.

Mrs. Joe Koesel: We had everything here that you could think of for entertainment, and the company kept it up and everything. The best swimming pool between Fort Worth and El Paso, the best hospital. We had a good school, grade school. A good church, it was a nondenominational church. We had a big clubhouse over here, we had a guest house over here, just everything. We had polo teams. We had baseball

teams. We had a nice golf course, the first golf course out in this country. And the company fixed it up.

Joe Koesel: Christmastime the company'd give every kid a nice present, over there in the big old theater, you know. Each kid, according to age, they all got something. Big old trucks, big dolls. I mean, they were nice presents.

Mrs. Koesel: Big dolls. And sets of dishes, they were large enough you could almost use them on the table. And every gift was that way. Basketballs, and footballs, and things like that.

Joe Koesel: Every Labor Day they'd feed sometimes six thousand. The whole county and everybody would come, you know. Everybody. Every Labor Day they'd have a big one. Then they'd have a baseball game, trapshooting, such as that.

Mrs. Koesel: When you wanted your house painted on the inside, you just called the painting bunch, and they came and painted.

Not all camps were cozy
communities: this bleak view
captures a camp near Wink
in 1930.

Joe Koesel: They painted one color, and that was white!

Mrs. Koesel: And when something went wrong with your plumbing, you called the plumbers. They came and fixed it. The company paid for everything. They had their own plumber, they had their own painters. They had the gang for everything. We paid two dollars a month for electricity. We had our own electric plant.

In the mid-forties, *Mrs. Robert Boykin* lived in a much smaller and simpler camp run by Standard Oil of California near Royalty:

The families were not very large; most were rather young people. Most of the people in the camp were Californians. I think we had about fifteen families. And each of us had our own little house. Sand went right through them, believe you me! But they were comfortable enough. They were wood-frame and would have at least two bedrooms.

We had to make our own fun. We'd have picnics and things of that kind. You couldn't get out at night without a light at your feet because of the rattlesnakes and those big black spider things, tarantulas. Scorpions in your shoes: automatically, to this day, I've been known to dump my shoes out before I put my foot in one.

It was ten miles to a loaf of bread. We went to Grandfalls, that was the nearest place we could shop. Monahans was our chief shopping center for clothing and for canned goods. The first time I saw the vegetables, I refused to buy any, they were so wilted. And the woman who had taken me in laughed and said, "You'll get over that!"

But this bunch of Californians got a bulldozer and scraped off a square block and planted their own garden. And in the morning you would have a basket of vegetables and melons and whatever that you didn't have to do a thing for. It was lovely.

Mr. and Mrs. Skeet McAuley moved from a three-room shotgun house into company housing in the Gulf camp near Monahans in 1948.

Mrs. Skeet McAuley: We got moved into a new house. It was real nice: three bedrooms and a bathroom. And we had a pretty yard. We were so happy to get a nice place. We lived there thirty-two years.

[The houses] were all white out there. They had a gang that painted the houses. There wasn't any carpet on the floor or anything like that. Hardwood floors. You'd wax them. In the kitchen and bathroom they had linoleum. We paid ten dollars a month for house rent. They paid the utilities, all but the telephone. We didn't have a telephone out there.

Oh, and we were all neighborly. I believe there was something like thirty-five families out there. Seemed like everybody was so friendly.

The larger the company camp, the more likely it was to offer many recreational opportunities to residents; even time off the job was thus given over to company activities. Many employees found company socials one of the pleasantest things about camp life.

Mr. and Mrs. E. W. Purdy lived in the Humble camp in Wink in the late forties:

Mrs. E. W. Purdy: Well, we probably never had any more fun in our whole life than living in that camp. Humble had a great rec hall out there, and it was available for all the employees. Once a month we had the biggest dance, the biggest party, that anybody ever had. And nobody was excluded. El Paso Natural had a camp there, Stanolind had a camp there, Sinclair had a camp there. Everybody came to those parties, and everybody had a good time. The drilling companies out of Kermit, everybody came to those parties. There was no status [discrimination] in those.

Newly planted trees leaf out along
the streets of the Big Lake Oil
Company camp at Texon in 1928.
The grandstand of the baseball field
is visible in the background, behind
the oil derrick.

E. W. Purdy: As a matter of fact, every district—we had eight district offices—every district had their own recreation club. And we'd go around to their places. Denver City or Odessa and all those. They'd have parties different weeks of the month. About once a year, they had this big barbecue for everybody in the camp. And they had the golf tournament once a year. And a softball game. I think at one time they had baseball; when we got there, they just had softball.

Mrs. Purdy: And we made some of the best friends you ever had in your life. They took care of you.

You made faster, closer friends [in camp]. They were also nosy. They were also gossipy. And they never minded to look you straight in the eye and say, "How much did you pay for that?" "Are you pregnant?" "Where were you last night?" Never, ever.

But on the other hand, they were also your friends. If anything happened, they were right there. You never lacked for help. In fact, I think some of the friends would do more in circumstances like that than maybe even family would.

Most employees found more to like than dislike about camp life. It was not, however, without its drawbacks. One of them was abiding by camp rules, which were numerous and strictly enforced in some places. Mr. and Mrs. C. E. "Steve" Cullum recalled rules at the Cabot camp in Wickett in the 1940s.

Mrs. C. E. Cullum: It [the camp] was beautiful. It had to be; we were required to keep it up. It was really just lovely.

Steve Cullum: There were just certain rules that everybody was compelled to abide by. It was kind of an oasis in the desert.

Mrs. Cullum: Well, we had to keep our yard up. You had house inspection every month, almost every month. Mostly for safety, it was from the safety department.

Steve Cullum: You weren't allowed to keep any livestock.

Remembered as a neighborly
community: the Humble Pipe Line
camp in Wink during the 1930s.

Mrs. Cullum: No dogs. No dogs at all. Well, things just had to really be kept up the way they should be.

Camp life also had its occasional spats and squabbles. *Theola Starkey* recalled a camp in Hobbs in the 1930s:

In a camp, somebody's mad at somebody else all the time. Or else the kids are out fighting, and they have the mamas get into it. I expect mine was one of the meanest ones that ever walked on two feet. He whipped some of them and got whipped. But I never would allow it to affect my friendships. Some of the mothers did, though. One of the mothers told me one time, said, "You got the meanest kid in camp." And I said, "I know it, but don't tell me about it!"

Anne Swendig remembered a difficult camp of the 1960s:

It was close enough to town that most of the people had friends in town. And the people out there did not get along well. They were very, very clannish. You quickly discovered that if you spoke to one, then half the camp wouldn't have anything to do with you because you spoke to this one. And then if you'd speak to that one, who didn't want to have anything to do with you 'cause you spoke to the first one, well, then the first bunch didn't want to have anything to do with you. I didn't realize what I was getting into, and we really didn't assimilate into that camp well at all.

Camp life also had its psychological tensions. Employees were always aware in camp of their place in the company hierarchy; their rank and standing at work determined the size and location of their house, as well as the friendships they made. The closeness of camp life could make up for the absence of kinfolks, but it also allowed little privacy and individuality. *Patience Blakeney Zellmer* reflected on camp life in the late 1940s:

The first row [of houses in the camp] is the superintendent, assistant superintendent, and then the section heads of the departments. They're on the first row. And then there's the gang pushers. Instead of mixing, there was this definite break of your station in the company. If you were in the first row or the second row or the third row.

Now, see, most of the roustabouts, they didn't qualify for a house. A gang pusher did. You had to be a gang pusher to get a house. Daddy was the inventory man, so he'd go out to the field maybe two or three times a week, maybe twice a week. The gang pushers, you know, they'd go out every day. Right next door to us was a gang pusher. We grew up with them, and we were the best of friends.

You know, we were taught you're as good as anybody else, but you're no better.

But then, the rest of the people that were roustabouts, and the yard men, the roughnecks, they didn't qualify for a house. They could buy a house or build a house in the lower camp. In the lower camp maybe there might be twenty, twenty-five houses. They would say, "I live in the lower camp." That in itself was discrimination.

Class distinction in that camp, I don't know, it was there. You felt it. You knew it was there as children. You knew that Mr. So-and-so was the superintendent, and his daughter got so, you know, she wouldn't speak to you. My mother became very unhappy. My daddy would say, "This is a good education, but I really don't want my children raised like this anymore."

Mother and Daddy . . . I mean, they were different. They came from a different background. They were different. That's the first thing I noticed, that we were different. That Mother and Daddy were different. They didn't just socialize with anyone; they sort of picked their friends, too. They felt different because of being from a different geographical area and religion.

So when I married, and I knew that I could leave camp life, I was relieved. I was tremendously happy to be free of that pressure. It was a pressure, that I know. And I didn't find it in school. I only found it in that camp. And my

Social functions were frequent in large camps: Joe Koesel (back to camera) receives an award at the annual Big Lake Oil Company barbecue on Labor Day, 1948.

mother found it in that camp. It's common in oil companies; it's common in the social order of oil companies.

It's a loss of privacy. You must realize that you weren't that far away from your neighbor, and your neighbor was maybe your boss, his wife. Everything you did was being observed. You felt like you had not a whole lot of privacy. So that limited you, too, in your friendships, because there was that pressure all the time. I think that really had more to do with it than whether people really liked each other and got along.

Late in the fifties, oil company management came to see company camps as expensive and outdated. In most Texas oil fields, better roads and better cars permitted workers to live in town and commute to work in the field; fewer workers wanted to live in camps. In time, as processes were automated, less production work had to be done out in the field. One by one, the company camps shut down. One of the largest camps closed was the Humble camp in Cisco. *Clell Reed* recalled:

That was a division headquarters, Cisco. Humble Oil and Refining Company built 140 houses. By the time I got there, the Humble Oil and Refining Company had sold all those houses but 32 to individuals, and they sold 32 to Humble Pipe Line Company [his employer].

We had a large, nice, five-room house with a shingled roof and beautiful oak trees, with water and gas and everything, with three dollars a month for rent. Electricity was two cents a kilowatt, and they furnished everything else.

After a while, the company was losing $1,000 a year on each house. From 1945 to 1956 we lived in that accommodation. Then they decided that they'd sell the houses. They gave consideration to people that were living in them, but you had to bid for your house, and anybody could bid on anybody else's house. We were fortunate enough that a friend of ours bid on our house and overbid it, and then sold it to us for what he bid for it, $2,800.

Few amenities were in place at the newly built Lion Oil Company camp near Snyder in 1949.

E. W. Purdy: The idea of the company camp to begin with was the fact that housing was just not available in these places. So they built these company houses. Then they did away with a number of these districts, consolidated, and moved the districts [offices] to towns that had housing. And they said, "Well, you'll have the opportunity to buy a house and get an equity built up in a place of your own." This was the reasoning that they put forth.

And, as I said, most of [the employees] were happy about it.

A fair sampling of local life: signs on
the outskirts of Hobbs, New Mexico,
in 1940.

7: Fitting In

The men, women, and children who followed the oil activity were used to being strangers. In each new town they first had to meet immediate needs like a place to live, and then they had to make new friends.

Sometimes making friends was easy. Oil field veterans who made each boom ran into the same people they had seen in the last oil field and kept up friendships as they moved from place to place. People who were transferred in the normal course of work for major oil companies often lived in company camps near friends they had worked with in other places. It was not hard to fit into the mobile oil community and to have a circle of friends entirely within that circle.

Fitting into the community outside oil was more difficult. Like rural people anywhere, farmers, ranchers, and small-town folks tended to be wary of newcomers. The discovery of oil on a farm a few miles from town brought a mixed reaction: delight at the expectation of a bonanza from black gold and dread at the thought of seeing a crowd of transient strangers descend on the neighborhood. Generations of sensation-seeking journalists told their readers that oil fields were wild, bloody places filled with con men, madams, and desperadoes; that distorted picture did not make it easier for oil field newcomers to gain acceptance. The presence of large numbers of drifters during the drilling boom supported the widely held image of "oil field trash." When the more conventional mobile workers and their families arrived, they, too, did not seem like "the right kind of people" because they lived in tents or shotgun houses but always seemed to own shiny new cars.

For the more settled oil field people, fitting into local communities was a gradual and occasionally painful process. Oil field workers who stayed in one place long enough to join churches, belong to clubs, or see sons on the high school football team usually found acceptance. Once local residents could see real economic gains from oil, in the form of royalty payments, fattened tax rolls, and better jobs, they included oil people in their social circles more readily. But all this took place gradually. By the time it happened, the boomers, drifters, and most mobile workers were long gone.

Mrs. Joe Koesel recalled the way some Big Lake residents acted toward oil field newcomers in the 1920s:

> Now, when we came out here to Big Lake,
> they—well, it wasn't very nice. [They treated us]
> like trash. The man, he had the drugstore over

there. There were two or three of we women sitting in the back of that drugstore. We heard him say he wished that damn trash would stay out of his place of business. He didn't appreciate it. We heard him say that. I never felt the same way about him after that.

Theola Starkey saw similar attitudes in the thirties:

Pearsall was a clannish little old town. They didn't like oil field people. They sure didn't like 'em.

[Fellow camp residents] used to laugh at me about getting invited to the bridge parties [in town]. And they wondered how I'd worked it. I said, "Some people have it, and some people don't."

But now, Monahans: that's the reason I love Monahans, it's the friendliest little old town! Everybody loves everybody else. Because the biggest part of them here are oil field people now. I love the oil field people.

Sometimes even decades of oil production did not break down barriers between oil people and other community residents, as *Mrs. Clell Reed* noticed in the late forties:

Now, Cisco was quite clannish. Most of all their people had lived there always. There was a lot of people there who never worked anywhere else but Cisco. Out in the West Texas oil field everybody was like everybody; there was no clannish feeling. But in Cisco, a lot of the people in town were a little bit, I would say, jealous of Humble people because they were better paid and had a better way of life than a lot of people in town.

Her husband's transfers by his oil company employer took *Mary M. Porter* to a number of towns in Texas and Oklahoma in the sixties:

Not very long after we moved to Nowata, Kathleen [her daughter] did become ill. I had only met one other woman whose husband also worked for Sinclair. I called her to get a referral to a doctor. She gave me the name of a doctor and said, "Be sure and tell the doctor's receptionist that your husband is an engineer, or they won't see you."

They didn't want pipeline workers or roust-abouts, people who did transient kind of work.

We were considered transients, certainly, not accepted into the mainstream of life. That sort of feeling carried over into church life as well. In fact, there were almost three standard questions in all the little towns: "What church do you belong to?" "What does your husband do?" and "Do you play bridge?" The most important, of course, was where you went to church. In Nowata, if you did not go to the First Baptist Church or the First Methodist Church, you were slightly suspect.

I can remember being told by a librarian in Sapulpa that I was a transient, and I couldn't have a library card unless a property owner in Sapulpa would cosign my application for a library card. I said, "Well, I rent property. Don't people who rent borrow books?" And she said, "Well, yes, but transients might take our books away. You just don't know what the public does to our books." And I said, "Does it not say Public Library out in front of this building?"

There is that terrible feeling of isolation. You move around a lot. In my experience, it takes six months to a year to become acquainted and feel integrated into the community. And in that time, if you need something, it's difficult. You have to leave friends and family.

I did find in Lindsay, Oklahoma, oil people were more a part of the community. But I found, especially in Lindsay, a drawing together of oil people. There was a young woman who lived down the street in Lindsay who had children the age of mine. And I commented to a neighbor one time that Doris did not really socialize with us. She didn't join our coffee klatsches and that sort of thing. Someone in the group said, "But Doris is from Lindsay. She doesn't need us. She has a sister. Her mother lives here. She doesn't need us. We need one another."

And I found that. We'd commiserate: "Well, how was that house you lived in?" and "Did you ever live in So-and-so?"

I think that people who had lived in places

like Lindsay all their lives did not want to risk knowing you and becoming friends with you. And I never understood it very much. I was always a little hurt by neighbors who waved at me as they went by but never came to my house to see me or never asked me to their house. I never quite understood it until we came to Midland, and we were here for eight years. Along about the third or fourth year, a couple of my neighbors moved away. And I had never had that experience. They were oil people; they were transferred away. For the first time I realized what it was like to have them move away from me. I had always said to my friends, "What are you yelling about? I'm the one that's going. I've got to go to a new town, I've got to unpack. Why are you crying? You're staying here." They never knew how long you were going to be there. I have been fortunate that in every town we've lived in—and we've lived in eleven—I have made at least one good friend who has remained my friend.

Separated from family and treated like outsiders in the places they moved to, oil field people depended on one another. They developed a sense of community and offered one another friendship and aid. By the twenties and thirties, neighborliness was a strong feature of oil field life.

Clell Reed remembered oil people in Crane in the early thirties:

> That was a close-knit, well-defined fraternity. There was a lot of love and devotion involved. People really loved each other in those days. They didn't just trust each other. They were all from somewhere else. And to come in and to know someone was a revelation of acquaintanceship that they enjoyed. It was good for them.

Cullen Akins lived in McCamey:

> We had a lot of friends there. Everybody was just like we were. They were there because the company sent them there. They came from Oklahoma, East Texas; they came from everywhere, and everybody wanted to get acquainted.

The breadwinner, ready to take work
where he could find it.

Will the next stop be the last one?
The worker and his young wife look
down a barren road in West Texas.

Only the youngest child shows no sign of worry as the family interrupts the search for work long enough to pose.

When disaster struck, oil field workers helped one another, as **Bessie Leonard** recalled:

We had boxed our tents up, but nevertheless, they burned. A man came over smoking a cigarette and threw it out the window and it went into shavings where we had shaved off a board that we had boxed up with. We'd just got it finished. And that's what caused the fire. So we was left without anything. The men in the fields went together and gave, made up purses to buy clothes and stuff, what we had to have.

It was one of those things, but we were pretty close-knit, because it seemed like one had to help the other one then. You wouldn't think we would. But you'd see someone who had a sick child, you know, or one of them was sick, why, they'd call for someone to come and help. Well, So-and-so would know what to do, so go over there. And it was a mouth-to-mouth thing.

Mr. and Mrs. Joe Koesel remembered Texon in the thirties.

Joe Koesel: Somebody'd get sick, they'd all pitch in and donate this and that and all. Of course, a lot of them would go and buy a car with the money you gave them, an old secondhand car or something. But they'd all pitch in, you know, when somebody'd get sick.

Mrs. Joe Koesel: I had a little girl die, and of course my first husband died. And when I came in from the funeral, why, the bosses' wives, two of them, they didn't go to the funeral; they stayed home and cooked supper for the whole gang. When I went in, the house was full of women. And that was the way it was everywhere. We'd all help each other, take care of each other. We took care of each other's kids. Everybody watched out. If I saw somebody's child out in the road, and it shouldn't be there, I went and got him out and told his mother. And that's the way everybody did. We took care of them.

Alice Keene had similar memories of Wink:

Some of your neighbors would help you. Well,

you'd help them. They always come in to help whatever way. And that is the practice yet today in Wink. Someone dies, why, everybody goes in. Different organizations, the church or something, will carry food. You know, they look after you.

In hard times, some oil field businessmen extended neighborly aid to customers who were down on their luck. **Ruth Godwin** recalled a grocer in Wink:

> During the Depression I think you got closer than ever, because everybody depended on everybody else. I remember one time a fellow went into a grocery store and held it up. Took some bread and milk and said, "Sorry to do this, but my family's hungry, and I'm going to take what they need." There were several in the store and they asked the store owner what he was going to do. He said, "If I knew where he was, I'd hunt him up, and I'd give him some more groceries." That was during the Depression.

"Red" Laughlin ran a garage in Wink:

> Old Red Fortram, he was in East Texas but didn't do no good. He come back out here, was working on a wildcat well out of Pyote. That was in the thirties. He had this Chrysler that needed work, so I put in new pistons and overhauled the goddamned engine and one thing and another. Made a pretty good car out of it.
>
> He finished that well out there. And he was supposed to have six wells to drill, see. And then they shut down. Red had his wife and three kids. He owed me about a hundred dollars on that goddamned car. He'd been waiting around, waiting around, waiting around, and finally he got his check.
>
> He come in there and he says, "Red, I got something to talk to you about."
>
> I says, "What's the matter, Red?"
>
> He says, "Well, I'll tell you. I don't know if they're going to drill any more wells over there or not. I have a hunch they're not, 'cause we didn't do any good on that, and I don't think they're going to do any more. I can set here. I got my check. I can set here and live on it, and pay

you, and hope that something or other shows up. Or I can not pay you and take that money I got and go down there to Arkansas where I've got a little farm, and we can put up enough food for me and the old lady and the kids to eat on for the rest of the year. I'm asking you, what would you do?"

I says, "I'd go to Arkansas." And that's what he done. He was a good guy.

Once local residents realized that all oil field people were not con men and cutthroats, workers and their families could begin to fit into communities. Socializing at the traditional rancher's barbecue to celebrate an oil discovery helped. *Alice Keene* went to such a barbecue in the twenties:

The first time I ever visited a ranch was to the Thorns' ranch. They had a barbecue. Mr. Thorn come around and asked me, "Don't you want some Son-of-a—?" He called it by this other name. And I thought, "Oh, my Lord, what in the world is this?" So I asked Fred, "What is— you know?" And he said, "It's known as Son-of-a-Gun, too." And I said, "Well, what is Son-of-a-Gun?" He said, "It's a kind of a stew," and he told me what all it was made of. Oh! Then up came Mr. Thorn with a bowl of it for me. I said, "I don't know whether I want that or not." He said, "It's good, I made it, it's good." Well, I tried it, and it was good: tasty and good!

After they had their barbecue, they went to dancing on the bare ground. We didn't dance, but we watched them dance. We all had a good time.

Mrs. John Berry remembered giving a barbecue in the fifties:

They came in the spring of '53. I believe it was on May 26 they brought in oil on the Dyers', the ranch west of us. Then they went right across the fence and brought in our first one. I couldn't believe it. They thought they hadn't brought it in, and it came roaring up at daylight the next morning. So we were just delighted!

In those days the custom was to have your oil barbecue. The ranchers furnished the beef, and the oil people furnished all the fixings, you know, beans and potatoes. So the Dyers had theirs in early June, and we had ours July the third of 1953. They were kidding me all the time. I told them, I said, "I'm not going to buy any beer; if you want beer, you'll buy it." But they furnished all the things. We had a new hay truck, a flatbed just freshly built. John put it by the windmill, and that was our table. It was full of food. We furnished four quarters of beef. And we had 450 people. We asked all the oil people and the neighbors. The ladies, some of them brought desserts, and some of the men helped John barbecue. It was delicious. The oil people strung lights from the windmill tower up through the trees. It was all lighted, and it was like a fiesta to look at it.

The oil people are just like any other people. There are all kinds. The drillers, the kiddos who worked, they were just like someone's neighbor boys. We got to be friends.

Going to church often helped newcomers fit in. **Mrs. W. W. Silk** came to Wichita Falls in 1912, when it was booming from North Texas oil discoveries:

My first impressions of coming to Texas are somewhat mixed with pathos and homesickness, because the country was so very different from the country from which I came, Alabama being full of trees and flowers and beautiful weather. When we landed here, we were in the midst of a three-year drought. My husband had told me that we were only coming for a few months, and when I realized that he had come to stay, I was a very disappointed woman, because there was nothing here that I was accustomed to. There were no trees, no flowers.

But one thing I found is a wonderful type of people. And that finally reconciled me to being here. Because, different from the old Southern customs of the Old South, they didn't wait to find out who my great grandparents were [but] accepted me first and let me prove who I was.

The second day that we were here, we attended the First Methodist Church and met some very prominent people who have been our friends for forty years. They took us in as if they knew that we were all right, and yet we had not given them the record of our ancestry. I remember our first pastor when we came here said in one of his services one day, "Turn to the right of you, and see if you know the person sitting next to you. If you do not, shake hands with them. Maybe they're just as lonely as you are." From that time on, I felt that I was a part of Wichita Falls. Then there were no barriers between strangers and newcomers—and home people.

Mary Rogers attended a little church at Joinerville in the early thirties:

Over there at Joinerville they had a little church, and they had a part-time preacher. He preached two Sundays a month there, one Sunday at another church, and the fourth Sunday in the month over at another place.

The first Sunday that we were down there, my husband had to work, because all the men had to work on Sunday. So the children and I went to Sunday school. We went over there, and there was quite a few people there. All the people who were there were oil field workers or oil field people. The door was locked, and we couldn't get in. The pastor lived right next door to the church, of course. We stood around there a little while, and somebody suggested, "Well, the preacher's gone and won't be here today, can we go on in and have a little Sunday school?" I believe one of the men went over to the pastor's house and asked the lady if she objected if we went in and just had a meeting. They got the key, and they couldn't open it [the church door], so they just had to break the lock, break the door, kind of. But anyhow, we had a good little service. We had some songs and prayer, read something in the Bible, and talked, and kind of found out who was who.

Well, that kind of opened the eyes of some of those natives. I don't know what they expected of us: all to carry guns and shoot 'em, steal everything we ate, or I don't know what! But, anyhow, they really accepted the oil field people.

Bill Briggs recalled attending church in West Texas in the early 1950s:

A great many of the towns, really, you had two levels of society, the oil people and the town people. I was fortunate in that I dealt with both. Once I became rather a permanent fixture here in Midland, I was accepted, again by both. You had to have been here awhile.

Of course, one of the ways I found to be accepted by the "establishment," it's just because of my nature I happen to be a churchgoer. I must admit that I used to play a little game, on occasion at least. And that would be to go to church the first free Sunday I was in a town. And I would invariably, no matter how full or empty the church was when I got there, walk down to the second row of pews and seat myself. And never look back: always forward. Invariably, when church was over, there'd be any number of people come to introduce themselves to me, just wondering who this stranger was. And it meant a good, free Sunday dinner every time! The real knack of it was to find out who had the prettiest daughter.

School activities could help newcomers fit in, as **Anne**

Swendig remembered:

When we left Hobbs, we went back to Seminole. I hadn't enjoyed Seminole the first time, and I didn't want to go back. And I went back to Seminole, and I had a marvelous time. We lived in town. It was a small town, and the Amerada people were probably the biggest group of non-farming people; it was the difference from going there the first time as "that new young engineer's wife" and going back the second time as "that new young athlete's mother." It was just open sesame to everything. Like a lot of small towns, athletics is very important and everybody follows the team. They made a lot of the boys, and I was their mother, and to heck with being the engineer's wife. I had a ball!

It was never easy to be new in town, but by the 1960s oil field people who were newcomers could expect to find acceptance much more easily than their fathers or grandfathers in oil. Fewer workers faced constant moving to keep jobs; more could stay in a place long enough to put down roots. In Texas, most places where there was oil activity were places where oil had been produced for some time;

Among rig builders, extended families commonly followed work in the oil fields. One of them, George W. Rumbaugh and four of his five rig builder sons, pose with their families at Batson in 1906.

local people were familiar with oil. Thus, the once-common insult "oil field trash" found its way onto bumper stickers, as a rallying cry for oil field pride.

Vera Lacefield: Back when we first started out, you were oil field trash. You were just oil field trash. But I don't think [they feel that way] now, you know. Years back they dropped that. You don't really hear so much of that.

Anne Swendig: I wonder if, way back when, they used to say the same thing about pioneers, the ones who went out and left their homes, and went out and chopped down the trees and made their homes. Every generation, the settled ones probably called the mavericks that went off and had nerve enough to make a new life, they probably called them the same thing.

8: Growing Up

By present-day standards, the oil field of the early twentieth century was no place for wives and children. Some wives refused to follow their husbands to remote oil leases; some wage earners thus chose to "batch it" in the oil patch, keeping their wives and children in separate households elsewhere. Most working people, however, could not bear the emotional and financial costs of breaking up their families. In order to stay together, they were willing to take on the challenge of bringing up children in rough surroundings. Where there was oil, there were usually swarms of children.

Children who were raised in the oil field did not grow up with many material luxuries. As mothers struggled to keep house under difficult conditions, children were often left to amuse themselves in bleak surroundings. Both teachers and doctors recall, however, that oil field children were usually healthy children; the relatively high wages their fathers earned bought nutritious food and sturdy clothing, even if the wind often whistled through the tents and shotgun houses in which they lived.

Like children anywhere, oil field children had their joys and their problems. Their problems were largely the result of frequent moves and separation from kin outside the immediate family. Those who grew up in the oil field often say that they think their childhood made them more adaptable and resilient, more self-sufficient and outgoing, than children who did not move with oil. They usually add, however, that their early life was not an easy one.

In the oil fields of the twenties, families who moved frequently could not be sure of having a doctor at hand when it was time for a baby to arrive. *Bessie Leonard* recalled helping at a birth in a McCamey tent in 1926:

> My mother and I, we delivered the first baby
> that was born in McCamey. That was before the
> hospital and doctors came in there. This woman
> was having trouble, and her husband kind of
> kept asking around town until he found some-
> body that could go. My mother and I went, and we
> did deliver, and it was a girl. We just gathered
> up what we could have, because we didn't know
> what they had. And we just took off in the car
> and went out there. Everything was OK.

Four decades later, most expectant mothers had access to modern medical care in the oil field, but the need to move with their husbands' work still presented unusual problems. *Anne Swendig* remembered a move from Hobbs, New Mexico, to Terry County, Texas:

I was expecting the baby any day. I had chosen to have the baby in Seminole, rather than in Hobbs, because I ran into a nurse and asked her about doctors and hospitals, and she was going to Seminole [to have her baby]. And I figured if it was good enough for her, it was good enough for me. It was a thirty-minute drive [to Seminole], and I'd had the last baby in twenty minutes. John was nervous.

Oh, I had waxed and polished that trailer house: I mean, I had it immaculate. And I went double-stamp shopping, and I got more groceries than usual so everything would be stocked when I had the baby. John came home from work and said, "Guess what? We're moving."

"When?"

"We're supposed to move to this oil field camp out at Adair; we're supposed to go out there and look at it tomorrow, and see about moving in."

I said, "Did you tell them that you're expecting a baby?"

He said, "Well, not exactly. I guess I looked kind of funny when they said that, and they said, 'Is something wrong?' and I said, 'Well, my wife's expecting a baby any day.' But they didn't say anything."

We went over to look at the camp on Thursday. They already had one trailer house out there, and they didn't have connections for a second one, so they said they'd appreciate it if we waited until Saturday, and they would get their roustabout crew working on it on Saturday, and we could move in. So we planned to move on Saturday.

On Saturday morning I woke up with labor pains. I wasn't really in labor, but I knew it wouldn't be too long. And John just kind of itched and twitched around. Finally I said, "Would you feel better if I went now?" And he said yes, he really would. So we woke up these neighbors of ours that we'd met since we'd been there. They kept the two boys, and we went to

Seminole. He left me off, and came back [to Hobbs], and was trying to get ready to move. Finally, he took the two boys over to Midland, where we had friends that said they would keep them while I was actually in the hospital. Poor guy, he drove seven hundred miles that weekend, just back and forth, between the hospital in Seminole and Hobbs and back, and bringing the boys to Midland and back, and moving the trailer house to Adair.

I had the baby Sunday morning, and he moved. They came in and asked for information for the birth certificate. They said, "What's your address?"

And I just kind of looked at them blankly and said, "Do you mean when I came in or when I go home or today?"

They said, "What do you mean?"

I said, "Yesterday I lived in Hobbs, New Mexico. When I go home, I'm going to an oil field camp up in Terry County, Texas. And if you look out the window, you might see our house go by now. So, I mean, take your choice."

They thought that over for a while, and finally they said, "Where was it parked at the time the baby was born?" Well, it was still on the New Mexico side of the line. So her [the baby's] birth certificate says New Mexico, although it's registered in Seminole, Texas.

John finally got moved in over there Monday. Tuesday morning he came over to see me, just exhausted. He said, "When are you coming home?"

Being smug and thinking, "Well, my doctor will never let me go home before three days," I said, "It's up to the doctor."

He said, "Well, where is he? Could I talk to him?" And I thought, "Oh, yes." He went and talked to him, and the doctor said, "Oh, that's right, your wife's a registered nurse, she knows how to take care of herself, yeah, she can go home now." I could have killed him! I really wasn't too enthused about this idea. What could I say? So we went home.

The trailer was out in the oil field camp, and they'd just gotten it hooked up. And all my cleaning: there was sand yea deep on the floor! The air conditioner wasn't in; there were flies all over the place. But everybody was so nice to us, they really were. Somebody was having a birthday party that afternoon, and we were invited. So we took the two-day-old baby, and all the kids played with her.

You know, when John went to get me at the hospital, he was hauling out all my "unpregnant" clothes and trying to find a dress that he thought maybe I'd like to wear. He dragged out the ironing board and the iron, and he was fixing to do this. There was an old Cajun from South Louisiana living there; he may have been a pumper, I don't know. He wouldn't have John iron that dress; he had to take that home and let his wife do this dress for me. So everybody was real nice.

Once babies were born, mothers and infants faced new problems: oil town records show that intestinal infections were the greatest hazards to infant survival, even as late as the 1950s. These infections, frequently epidemic where water and sewage disposal systems were inadequate, spread by means of polluted water and swarms of flies; both were common in new oil settlements.

Dr. D. W. Davis practiced in Beaumont in 1902:

At that time they were losing a great number of children with what we called *cholera infanteum* [infectious diarrhea]. I worked almost night and day with those children. Some of them I'd take when they were a living skeleton, almost. And I didn't know much about drugs or anything of the kind at that time because I'd—all I'd gotten about medicine, really and truly, at that time, was what I had studied in the medical books, and I hadn't had time to do a great deal of that. But I learned how to save these patients. Simply castor oil, one big dose. After that, I gave them ten drops of castor oil every two hours, and worked their bowels out, and gave them osteopathic treatment. And some of the most prominent men in Beaumont today [the 1950s] owe their life to that treatment.

Virginia Hoffman cared for her young family in booming Ranger in 1917:

We rented a little two-room cottage, the funniest-looking little place. I had a little six-month-old baby, and these two little children. So I had all I could do. The baby, we kept the baby in a wooden box. And Norman [the toddler] was determined to put the baby's eyes out. We had to watch him so closely; he was so interested in the baby, how the baby's eyes would move, that we had to watch him to keep him from putting the baby's eyes out.

Well, the baby got sick, and we had the doctor. The doctor gave that little six-month-old baby a dose of calomel. And it nearly killed the baby. So I picked the baby up and left the children with Frank [her husband]; I took the baby to Fort Worth and put it in the baby hospital. They had a baby hospital there. But the baby only lived a week in the hospital. The hospital said that we got him [there] too late. But it passed away. My baby is buried in Fort Worth.

Oh, I'm sure if we hadn't gone to Ranger, why, we could have reared that baby also. I know that if we had been living in the normal conditions that I had been living in up to the time I left Dallas, I know that baby would have lived, because it would have had every attention, you see. And you couldn't have every attention there [in Ranger], because they didn't have it. I guess the few doctors that were there were just people that were after the dollar. That was all.

Bessie Leonard recalled McCamey in the 1920s:

I can only remember one baby that died. Just died of malnutrition, really. The mother couldn't nurse it, and it had a hard time feeding. But they have things today like crib deaths. We never heard of that. There wasn't any such thing. They slept with the mother and father. You didn't have a separate bedroom for each child. If it got cold in the night, your baby didn't lay uncovered.

When this thing come in that babies slept off in another room, we lost more of them than we ever did when we was living right out in tents in the open.

Little is known about child abuse and neglect in the oil field. Crane resident **W. A. Allman** remembered one incident:

This man was pumping for me on the Duell lease out here, and his wife gave birth to a little boy. They had four children then, and the baby made five. He wouldn't take care of her. She was out there in a little shack on the lease, and the lady developed milk leg, and she died. He decided to hire somebody to take care of them and just live out there with them, but the son of a gun, every time he got a paycheck, he would come down, hit the line, and get in one of those whorehouses, and drink, and spend his money, and he wouldn't take care of his family.

I got tired of it. I went out there, and I saw how these little children were. The baby was about to die. He had his mouth full of bread. It scared me. The lady [hired to take care of the children] wanted to quit.

So I brought the children over to my house. I didn't ask him anything about it. I just brought them in. Some other people helped us, and it wasn't long until we had them distributed around to people in Crane. He was in an old whorehouse, and I went in there and got him out and just beat the socks off him, put him back in the company car, and made him go to work.

He was irresponsible because he drank. He would get his money and go drink and then he wouldn't take care of his family.

Dr. L. Rose Robinson, a pediatrician in Gladewater and Kermit, treated many children of oil field parents:

We were astounded at how well they looked after their families. For the most part they'd come in with a child in their arms to the clinic. They never called you to the home very often. And of course, we didn't go by appointment; we just went straight down the list, because telephones weren't plentiful in the oil fields. And they'd bring their children in for care. They were well fed, well developed. The fact is, the oil fields at that time [the thirties] was the only place where there was enough money to furnish good things for the children, food and clothing. Now, they lived ordinarily in oil field shanties—as we called them, shotgun buildings.

Of course, they move so often, and the education of a child might suffer from the moving. But for the most part in Texas, the money is where the oil is, and so the educational facilities are often better [in school districts with oil production] than in other parts of Texas.

They look after their children. They're not neglectful of their families.

Parents, however, could do little about the barren surroundings that were often an inescapable part of daily life out in the field. *Mrs. W. A. Allman* remembered bringing her family to Crane in the 1920s:

Our children were real small when we first came here, and we lived at the Savoy Hotel for two weeks. It was terrible. Our children had no place to play, and any time you would stick your head out of the door, somebody would holler, "Hi, baby!" We would have to leave the hotel and go downtown to a restaurant to eat, then back again. The rooms were very small, and crude as they could be. There was no air conditioning at the time; we were there during the summer, so we suffered from the heat.

[Then Mr. Allman found them a house.] It was just a two-room house, and it had a small clothes closet built in the corner. It had a drop light in each room. And no sanitary facilities at all: we went out to the alley to go to the toilet. Sheetrock on the inside and wood on the outside, with a tin roof. We didn't have any lawn at all. No water and no trees. My children were small, and they weren't used to that kind of living. We kept them in the house as much as we could.

They found room for two new
puppies in this East Texas tent of the 1930s.

Life was as bleak in the trailer court in which Martha Lyle and her children, Karolyn Hendrix and Tom Wilmeth, found themselves in Odessa in the early 1950s:

Martha Lyle: The trailers were pretty close together in the courts. There was never any privacy, no yard. That always hurt my feelings, because the children didn't have anyplace to really call their own. They could take a toy out to play with and lay it down for just a minute, and it'd be gone. The children couldn't keep anything. Roger played with his toys one minute, turned around, and they'd be gone. You couldn't keep anything.

Tom Wilmeth: We played in and out between the trailers.

Karolyn Hendrix: And one time Terry, our youngest brother, was outside early in the morning when a car of sleepy roughnecks came in and ran over him. It was real, real sandy, and somehow it didn't hurt him at all.

Martha Lyle: They just backed over him; he was under the car, you know.

Karolyn Hendrix: Then, another time, Mother was sick with the flu or something, and I was supposed to be watching Terry. He got out on Highway 80 there in Odessa, right out of my sight. All of a sudden I heard all these cars, and it was a three-car pileup, because that baby was in the highway. He had two near misses right there.

Martha Lyle: We did have some nice neighbors when we lived in a trailer. Met some of the nicest people. Everybody was in and out of everybody's trailer all day long. I used to have a friend who would drink coffee and sit all day.

But you know, living in a trailer camp with children, there were children everywhere. If you didn't want to associate with anybody, your children brought people in anyway. It was impossible not to be that way. It was impossible to be away from the children or have any time to yourself. Still, living in a trailer meant we could live together as a family. I could have stayed in one

A new car and a worn bedstead frame
a ragtown family in Depression-era Andrews.

place, and Daddy could have gone off to work and left us in one place, but I wanted us to be together. I couldn't stand it any other way.

In the absence of playgrounds and more organized activities, children roamed at large, watching adults at work, playing on oil field equipment, doing odd jobs, and getting underfoot in town. **Plummer Barfield** recalled growing up in Batson during its boom days in the beginning of the twentieth century:

> Kids didn't fare so rough. People didn't pay no attention to kids, 'cause a kid was considered a kid until he was grown in those days. I associated with men ever since I was twelve years old. They didn't care what they said in my presence, or what they done. I seen lots of things that it wouldn't be safe to tell. But I wasn't in any danger. The only way a kid would have ever got hurt would have been a stray bullet or a horse run over him, something like that. If he had sense enough to keep his mouth shut, you could be right in the middle of it. You're just like a dog standing out there. They don't pay no attention to you.

Skeet McAuley lived in booming Electra in 1911:

> Well, about the only thing we did was kick-the-can and—what'd they call it?—what you did was run across from one side to the other, and somebody'd try to stop you. I forget what they called it now. Anyway, that's about what we did.
>
> I worked for a meat market. There were two different ones. One of 'em was named Tough Steak Meat Market. They had that on that sign, a big bull on the window, painted on the window, and it says, "Tough Steak Meat Market: We Put Up What We Advertise." They paid me a dollar a day. That was six dollars a week when school was on. And I'd get up in the morning and go down there before school and work till eight o'clock; I'd go after school, work till nine. I made deliveries for them. I was about ten years old.

V. L. Cox: When I was young, real young, growing up, we'd climb those old wooden derricks up there at home. One time we swiped a .22 automatic off a punch board. One of us would

A small businessman peddles the
Sunday paper in Borger, late 1930s.

be on the ground with the gun shooting at the rest of us up in the top of that old derrick. And how we kept from getting killed, I don't have any idea.

Karolyn Hendrix remembered two favorite diversions of many oil field children:

When we were kids, it was a real treat to go out where they were drilling, go out and take supper out to the guys. We got to play on the floor of the drilling rig. That would be unheard-of now. But it was fun.

We'd go out and hunt jackrabbits. And at night, it was unbelievable. There were lights everywhere, drilling rigs everywhere. Every forty acres there was a drilling rig.

Charles Stroder and Patience Blakeney Zellmer both grew up in large company camps:

Charles Stroder: So many people out there [in West Texas during the 1930s] moved from East Texas. There I'd grown up in a little old nice community in East Texas where I was kin

to most, all my cousins and all in school. Little creek behind the school. Moving out to Forsan the first year, that was pretty much of an experience. It was quite a bit different!

[The Stroder family moved to the Lion Oil camp near Wink.] An oil camp was one of the best places to grow up. It had a lot of people, just very close. Our dads worked together. We were all very close.

They didn't have a community house. But we'd get into the big garages and play when it got too hot to play outside. We'd move into the truck garages and things. You know, they closed the swimming pools a lot in those days. They knew that it had something to do with polio; they thought so. We'd find some old saltwater pit out there that we'd swim in, half oil, half salt water. But it was all right. Then we had a good library. We read quite a bit, or I did.

Marble games were a big thing. There were about three or four camps back in there, and we'd challenge them to a football game or a marble

Four of Mr. and Mrs. Skeet
McAuley's children pose happily in
their Sunday best.

game. They'd come, and we'd try to win all their marbles. And we were playing for keeps.

We hung around together. Rubber guns was one of the big games. And slingshots. We hunted a lot with them, and we were good! We'd kill rabbits and all. We'd trap in the wintertime; never very successfully, but it was fun, I guess. And we'd just roam around the country. It was probably as much outdoors as if we'd been living over in the mountains somewhere.

Now the oil companies have more barbecues and things. We didn't have them; [the camp] wasn't large enough, I guess. We'd have maybe a picnic.

They had about three or four movies in Wink. And some of them were pretty bad. I remember one. We'd ride our bicycles or our mothers would take us in on a Saturday afternoon. I still remember seeing a movie, and all at once there was a naked woman. That was back in the late thirties; when I think of that, that kind of surprises me. Sure was a topic of conversation around our cir-cuit for a while. We didn't exactly understand all of it, but we'd understood the picture there. I guess we went to the wrong movie or something. I remember it was one of the movies on the back street. Instead of going to Tom Mix, maybe we'd gone to something else.

Patience Blakeney Zellmer: We had freedoms! We were mischiefs! Let me tell you some of the things we did. Us four younger ones, we would go down the railroad tracks. There was a water tower on the tracks, we'd climb that thing, get to the top of it. Somebody in the camp would see us: "There are those Blakeney kids!" And they'd have to come and get us down. We finally found the way to get into the company warehouse where all the furniture was stored. We'd go in the afternoons. They had a piano there, and we'd get the old clothes on. We'd just stay there all afternoon.

We had a delightful childhood, when you stop to think. You know, children are so organized today. We weren't organized. We found plenty to do.

Traveling circuses, shows, and carnivals brought new sights to oil towns: two Snyder boys study an elephant in the early 1950s.

Oh, that sand! I can remember the sand blowing so hard. We'd be out playing. There was a baseball field just east of the house, and we'd go and play on those stands. My mother and daddy would come after us little kids; they'd have to put diapers around our faces and lead us home. The sand would come up just all of a sudden, see, and we wouldn't pay any attention, and all of a sudden you couldn't see two feet in front of you. They always had to round us up.

Daddy was one of these parents that gave us things. We had baby lambs, we had goats, we had chickens. We could have them there in the camp. We had books. And we had music: every Sunday afternoon, he would make us come in and sit down, and we would listen to the opera or a concert. He did teach us to love things like that. He spent a lot of time with us.

[They moved to the Humble camp outside Wink in 1942.] We had a baseball team. We had tennis courts. And of course they had what we called the rec hall, the recreation hall there in the camp. They had a jukebox; we just danced, danced, danced. They had a playground. And they had company barbecues, which were a lot of fun.

Across the road was the Stanolind camp. Right down the way there was a Pure Oil camp, and there was an El Paso Natural camp. The Shell Oil camp, they had the swimming pool. We had the tennis courts, they had the swimming pool. We had friends in every camp. And it was like a great big family in a way, because you felt like everybody had the same type of job. Every father was with the oil companies.

V-J Day, it was the biggest thing that ever happened in our lives, after Pearl Harbor Day. Of course, everybody was just elated when the news came. The kids were all kind of milling around in the playground in camp, where the swings and stuff were. Everybody was out, everybody was roaming. The kids decided, "We want to celebrate." And somebody had this bright idea that we ought to have a bonfire. Well, below the camp was a big pit [where] they burned stuff, all old equipment and junk, garbage, trash.

Someone said, "Well, let's start collecting; let's have a bonfire in this pit."

That was early in the morning. And I'll tell you, we collected all day long. Tires, I mean everything that would burn. I think there must have been fifty kids involved—everybody, all ages. I think it was about six o'clock; I don't know who set the fire off. It got out of hand, it was just roaring. It was going to travel. Somebody decided, "Well, we need to call the fire department; let's get this real exciting." They called the fire department. Then they said, "You better run and hide, everybody hide!" There were a lot of hedges in that camp. We got in the hedge or behind the hedge.

And then, all the men [in the camp], all of a sudden they realized, "We've got a big fire in the camp." Everybody runs down there, a big commotion, fire bells going off and everything. Of course, everybody just got scared then.

And here comes that fire truck. Well, those guys had been drinking all day, and they were just as drunk as skunks! The truck came weaving down that road, and they're barely hanging on. The truck upset, and the firemen, they had just fallen off that truck. And it was funny, and it was awful!

The camp superintendent, he was just about to pop! He threw one fit. They finally got the fire out. He told those men, "You find out who did that; your job's on the line," I guess. We stayed hid for a long time. We didn't know what was going to happen. I remember my father calling us home: oh, he was mad! Well, there was five of us. He lined us all up, and he said, "Did you do that?" He'd point at my sister, and he'd point at my brother Pete: "Did you do that?" No, nobody would say a word. Nobody ever said a word. None of the kids would rat on each other. Code of honor.

We had the run of the streets. My biggest adventure was to get up and go see what was going on in town. And I'd make all my stops. By the time we were nine and ten, we knew where the bad part of town was; we knew what to avoid.

But we might scurry around there to see what was there, what was going on. We would run over there to the printing office and watch the paper being printed. And we'd go down and stand by the saloon and see who was in there. I had friends that I would visit. We just had the run of the town.

I never felt—I never had any fear. At night a lot of times after football games, I wouldn't have a ride home when I was like thirteen or fourteen. I'd walk home. That was three miles, late at night. Didn't bother me a bit. You just had a toughness there, you know. We had to be self-reliant.

Mrs. Jack Rapp and her four sons
show off the boys' dog and pony and
Papa's nitroglycerin wagon in
Ranger, circa 1918.

9: Fighting Through the Grades

When oil field families moved from place to place, children never knew what their next school would be like. They might find themselves in modern classrooms with up-to-date furniture or in Sheetrock sheds without indoor plumbing. Sometimes there were not enough classrooms to go around, and children went to classes held in churches and storefronts, or outdoors. Wherever they went, however, mobile children faced two challenges: being the "new kids" in their grades and doing the work their new teachers assigned. If they were lucky, the school bullies would ignore them and their new classmates would be behind those in their former schools.

Teaching in oil field schools also posed special problems. Teachers, particularly in primary grades, often faced overcrowded classes on the first day of school. When a boom was on and school enrollments skyrocketed, there were not enough seats, books, or materials for everyone. Students in any grade were likely to be working at many different skill levels at the beginning of the school year, and those who entered in the course of the year were likely to have fallen behind their grades. It was not easy for teachers in overcrowded classes to give mobile children the individual attention they needed to catch up. And however much a teacher tried to help, her best efforts would end with the student's next move. Many oil field schoolmarms thus missed the most basic satisfaction of teaching, that of seeing a pupil's long-term accomplishment.

Classroom problems were not the only challenges faced by oil field teachers. Like other oil field residents, teachers put up with rugged living conditions; like others, most were far from their hometowns and kinfolk. Many an oil field school board, moreover, assumed the right to regulate teachers' conduct outside the classroom; in some oil towns, merely being seen at a roadhouse dance could cost a teacher her job, as could spending too many weekends out of town. Oil towns might be rowdy, but their teachers were supposed to walk the narrow path of strict virtue. Only high salaries made all this tolerable.

Once oil development fattened local tax rolls, oil field schools were transformed. Sheetrock buildings were replaced by ultramodern, multifloored brick structures. Dirt playgrounds gave way to floodlighted football stadiums. School districts that once lacked money for indoor plumbing installed air conditioning and built tennis courts. Though a sudden burst of oil development could still result in overcrowding, by the 1950s most oil field schools in Texas were modern and comfortable. In this respect, children in oil field schools were better off than children in many other places.

Two former students remembered Eastland County schools in 1919.

Carl Angstadt: We had it pretty tough for a while. No place to live, you see, no streets, and mud. The schools were bad, the teachers indifferent. Here in Eastland they had one school building. They built some old wooden temporary buildings. No seats, just a lot of benches to sit on. No playgrounds.

Theola Starkey: It was crowded. Just all kinds of kids in there, and everything else. It was good for the kids. If you weren't an outgoing person, you learned how to be one.

Ruth Godwin recalled school in Wink in 1928:

The first school building was a two-room building that was also used as a church. It was a frame structure. In 1928 they had to put up some additional buildings. They were frame buildings with Sheetrock inside and no outside walls; just a single wall of Sheetrock on the inside. They had a series of about four of those buildings, with about three rooms to each one, kind of like barracks buildings. There were about forty or fifty in my class at that time.

The teachers couldn't stand it. They came and went faster than the worst people. It was just rough. They had some students that were rough. I remember one time in my class the teacher went to give a student a whipping. These barracks buildings, at the end of each one was a cloakroom. She took him back there and was going to give him a whipping, and she carried a paddle back there. He was a big boy. He pushed her around for a while. He didn't hurt her or anything, just frustrated her. He finally got tired of it and pushed the Sheetrock wall out of the side of the building and took off. That was the last time I ever remember seeing him in school.

Children of roughnecks and roustabouts pose with their teachers for the 1929 Santa Rita School picture.

Mary Rogers described what her children faced in East Texas in 1931:

> [Joinerville] was another place they had no schools. They had only a four- or five-room schoolhouse, but they had no place for the children to go to school. They had a few tents around that they held classes in; they'd keep the rain out. They just threw up kind of sheds and places for the children to go to school in, and started building a nice school, which they did.
>
> We lived down there at Joinerville three years, and then we were transferred in the north field to a place called White Oak. They had a small school and had the same conditions as the rest of the country had had. Anyway, they just began to build there at White Oak, and they never quit. They have a wonderful school for their children now. I'm telling you, they just have everything in the world for their children. They had plenty of money; they built the best for the children. They furnished all the band equipment, all the football togs. Everything like that was furnished, everything.

Three teachers recalled what it was like to teach school in West Texas shortly after the discovery of oil.

> *Helena Grant*: In 1925, when I went there [Odessa], they were just beginning to smell oil in Ector County. People began to move in with their small children. Of course, my room, being the first grade, filled up first. The school kept filling up and filling up and filling up. They moved me down to the courthouse. We used for heating a potbellied stove with coal. When I first went down there, I had only about 30, but they kept coming and coming, until one year I enrolled 145. Not all at one time, but they came and went. They had built a [new] school by that time. And they had little chairs in there. Well, half of them would sit in the chairs awhile; then they'd sit on the floor and let the other half have the chairs. I've had as many as 45 little first-graders at one time. I don't know how I lived to be ninety years old!
>
> They were healthy. I wouldn't say well behaved, because they had been dragged from pillar to post, living in tents. They were behind, of

course, especially from moving around. We just managed to do the best we could for them. I always loved teaching, and I love children. I always took pride in teaching my children the best I knew.

I taught eleven years in Odessa.

Gertrude B. Fleming: While I was there [in Iraan in 1928], the school board offered me a job. They said, "We will pay you $135 a month for nine months."

Oh, it was wonderful! When I told those other students down there [at Tarleton College] that I was going to get $135 a month, they didn't believe me. The teachers probably at Tarleton College weren't getting any more than that.

I ended up, then, teaching fourth grade. And we had a building that had been built just for, well, I don't know how many students they had had that year before. The year before was the first year that they had had school there. The town was growing by leaps and bounds, so they

had to have more room. I didn't get a room to teach in. I had to teach in the First Baptist Church. And all that I had was a blackboard. A blackboard. We had no desks; the children sat in these divided seats, separate seats. I had a roomful.

Seems like I taught till sometime after Christmas or in the spring, most of the year I taught in that Baptist church. The other grades were all scattered out like that, too. I think both first grades and maybe the second grades they put in this old building. We called it the old building. It was a stucco building, a nice modern building, with bathrooms and everything. They had a cafeteria—I say cafeteria, well, we could go and eat there, anyway.

They had a problem in Iraan getting teachers because it was so isolated. Well, I made it fine there for a while, and then the boom was over, people started moving out. In '32, then, I quit, and I got married.

Allie V. Scott: McCamey was a big booming town. We had 1,080 children enrolled at one

The annual class picture at the Laird
Hill School in East Texas, in 1931.

time; say it was from '29 to '32, in those four years, because enrollment remained like that for three or four years.

The first year I taught here, it was over in a tin building, Westover. I enrolled 96 children. There was a big turnover in all the population and especially schools. You'd be lucky, as I did there, to have 10 children remain out of that 96. I enrolled 96, and 10 of those stayed here to graduate. One still lives here. Now, that shows how much turnover there was in the school.

We had five first-grade teachers. A minimum enrollment [per class] was 40 to 45; that was how many desks we had in there. And you'd walk down between them, the desks were so close together, your hips would hit on either side.

We didn't have a problem with sick children. Most of them were rather adaptable, even at an early age, because they had moved a lot. They'd come from out of state and so forth. We had them from California and all other places. If the mothers could, they'd help children, between times, to adjust and taught a lot of them at home. You just had to give them all you could, for the time they were here.

Keeping up with studies was not the only problem mobile children faced, as Pyote teacher *Loreet Loftin* pointed out:

They were a discipline problem. But you had to sympathize with them, because out on the schoolground, or on the way to school, or on the way home from school—everyone walked in those days—they had to fight the town bully to establish themselves. And you had to sympathize with them.

One mother came up and said her little boy, [who] was new there, the other boys in the fifth grade brought him to task, and they thought he was going to fight. Well, he wouldn't fight. He wasn't brought up that way. His mother was a widow woman and came there to live with her sister who owned a hotel in town. He was a great big boy, and the boys who were tackling him were small boys, and good boys at school, but you know what takes place when they get out of school.

I told this mother—this is the only time I ever said this to a mother—I said, "There's not but one way to deal with this little boy that's aggravating your son on the way home from school. He's a pretty good-sized boy. And I've never advocated that sort of thing, but if he would just knock that little boy flat. . . ." But her little boy wouldn't fight.

So she came to the principal, and she told the principal what I'd said. I thought I was going to lose my job. I told him I'd done everything I could. I'd kept one of them in after school, I'd kept the other one in after school, just on and on and on. I said, "There's not but one way to stop it, and that's for the new boy just to show his brawn and really take him." I said, "What are you going to do about it?"

He said, "I'm really going to spank the little boy that's been doing all this to him."

I said, "That won't do a bit of good. You'll give him a good whipping, and tomorrow afternoon he'll be worse than he was this afternoon."

He said, "I just don't think that's going to happen."

Finally the lady, his mother, just picked up and moved somewhere else. Said she wasn't going to bring her boy up in an oil field town like that, with all that sort of thing going on. She'd come from Tulsa or somewhere. But that's just an example.

We tried to protect them as much as we could. They'd sometimes have to buy their way, candy bars and that sort of thing. About the time they would feel more or less at home, and the other children would leave them alone, then the father would move on. About three months to complete a well, and the father was a driller or a tool dresser, and they'd move on. Same old thing over again.

They were very aggressive. They bluffed their way through because they didn't have a foundation. But I felt sorry for them. We tried to protect them.

Vera Lacefield and her daughter-in-law Anne Swendig noted that Mrs. Lacefield's son John did his share of fighting during his school days.

Vera Lacefield: One time I had to move. John, when he was little, he liked to fight. We didn't stay but two weeks, and the lady [from whom they rented rooms], she had a boy the same age as John. Him and John didn't get along too good, so she asked us to move. 'Cause her little boy couldn't fight. John had so many buttons pulled off his shirt. So many times I'd have to reinforce underneath, you know, to sew some more buttons. The only way he could get along was to fight. New kid.

Anne Swendig: That's true. Every time they move, they're always looking at the new kid. John said, "Well, first time somebody came up and said hello, I busted him." His mother was talking about him fighting so much in Louisiana, and he said, "Well, look at the size of those Cajun families: when you busted one, the next day you had to fight his brother, next day he had another brother, another brother, and the cousins came along before you got through. And after you fought them all, or at least put up a good show, well, then they became friends with you. That's the only way you got along."

When John went in the service, they were investigating his education, using the Army security agency. They gave him a blank with five lines to list all the places he'd lived and when, because they had to go back and investigate. We figure they're probably still investigating, if they really did, or it'd take them that long, because John gave up on it. He sent the form home to his mother, and it took her two weeks. She finally came up with forty-three different town names and twenty-eight different school names, and some of the twenty-eight he'd gone to more than once.

Vera Lacefield: I knew when school was out, he was supposed to be home. About ten minutes after time for him to be there, I thought, "Well, he's having a fight." I guess he did that till he was about eight or ten years old. I began to think he was going to be a boxer. But John never did manage to get hurt. Only buttons.

Once children reached high school, aggressiveness expressed in schoolyard fights could safely be channeled into high school football. In Texas oil field towns, high school football season was the most exciting time of the year, a time when the whole town turned out to cheer on the home team. No matter how new you were in school, you were accepted if you could play football or take part in its allied activities of band and pep squad. And no matter how new you were in town, you were welcome at the Friday night game.

Charles Stroder played on the Wink football team in the 1940s:

> The school was a center. Football was basic. Everything revolved around football and the school. Everybody in Wink played football, I guess, if you had two legs. We had good teams, good coaches.

Patience Blakeney Zellmer was a Wink high school student in the late forties:

> Of course, the football games were just wild. About a day or two before a football game, it would just be riots almost in school, just cheering. One class would start cheering, and then the other one, and it would go on for hours. We were just so up.
>
> Then they had a bonfire the night before the game, and a snake dance. The whole town turned out for that. And it just got to be feverish. In fact, they still talk, there's nobody like Wink that had the spirit. You would be so up for days. And then when you got to the game, it was just wild, just wild. The whole season was like that. The stands would be full.
>
> I know a lot of the times when we won a game, the feeling would be so high, so harsh, we knew that we were to leave that field the minute it was over, get out of that town. You had better leave town, because several of the boys would be

beat up, cars would be run off the road. It was feverish.

Robert Horn had similar recollections of football mania in McCamey:

> They had quite a football town. They played on a dirt field. They went out and got just about anybody they thought was big enough to win, and they played football. They beat everybody in the country.

> If I remember correctly, San Angelo was the favorite target of the McCamey Badgers. They played San Angelo one year and beat them. After the game, they were staying in the hotel, and they just literally wrecked the hotel and wrecked the dressing rooms and everything else.

Bill Beckham of Kermit reflected on football's enduring popularity:

> I think a lot of it is the competitive spirit in West Texas. I guess it's sort of like horse racing in some places. They all go: roughnecks, rou-stabouts, and everybody else, all go to the football game on Friday night. Of course there's a lot of wagering on the side. Not supposed to do it in the state, but there's a lot of it goes on. I think that [helps make football popular], and the competitive spirit, plus the fact that people out here just love football. Some of these towns, you can't get a crowd for a basketball game, but a football game, they'll sit out.

Siblings Karolyn Hendrix and Tom Wilmeth lived in many oil communities during their childhood. They offered contrasting views of their experiences.

> **Karolyn Hendrix**: You really had to work to adapt to all those changes. I knew I just had to get by more on personality than on what I did. I could draw. I had real artistic ability, and most of the teachers in the grades would recognize that. So I'd always be working on some kind of project like that so I didn't have to do what everybody else did.

> I think that moving around as we had to was a handicap. I didn't realize it was at the time, but

it was a handicap in education. To this day, if I'm studying, I always have this feeling, well, I'm really going to get into this when I have the time. See what I mean?

Tom Wilmeth: One thing moving around when I was growing up gave me, it gave me the ability to meet a complete stranger and feel very comfortable around him. I meet the public, and I meet strangers every day. Living like that prepared me to do that. So, in a way, moving around prepared me to have a good life. We were educated in human relations. That's the main thing that schools fail to teach. They teach everything except how to get along with people. And if you can't get along with people, you can't do anything. We may have been deprived in a lot of things, but while we were going through those times, maybe we were being a little more prepared to face the world than kids that stayed in the same place.

10: Good Clean Fun

Work took up most of the day for men and women alike in the oil fields. Until the Depression, shifts were commonly twelve hours long on drilling rigs. During booms, rig builders, pipeliners, and other workers often put in more than eight hours a day on the job, as did waitresses and salesladies. Women who kept house and raised children under rugged conditions worked just as hard.

Time did not lie heavy on the hands of adults in the oil patch, but they had some free moments for recreation and leisure. Drilling and construction personnel were frequently idle between jobs; teamsters, truckers, and rig builders had a varied schedule. Wives and mothers found a few minutes for coffee with friends and turned shopping trips into social occasions. Athletic events always drew crowds, and movies and traveling shows turned up even in isolated areas.

For some workers, horseplay and brawling were important ways to let off steam. Physical courage and endurance were respected among blue-collar workers; the man who fought the longest and hardest was top hand, off the job. Sometimes the fights were waged between different work groups, as when rig builders battled tank builders and pipeliners. Other brawls were repeat engagements between old friends or enemies. When life was too quiet, there was nothing like a good fight to top off a Saturday night on the town!

Harry R. Paramore recalled lively nights in Beaumont:

> They usually would be somebody that would have a disagreement, and they'd have to step outside. I remember one time I went downstairs to the store; a couple got into an argument up there. As I stepped out the door, why, a bullet come down through the floor right in front of me. They used to have some knock-down-drag-out fights there. I remember one time in particular, there was a general fight up there, and one fellow got knocked down. His wife said, "Get up, honey."
> "Hmph! What's the use? They'll just knock me down again." Thought he'd better lay still.

As *Bill Bryant* told it, fighting was just part of the routine in Sour Lake:

> We would go down to the slum and get a few drinks, and we'd go back to the dance hall and we'd dance awhile. They got twelve and a half

A left hook meets a jaw as two
members of a Lamesa seismograph
crew mix it up in a boxing match in 1929.

cents for that. Then we'd go to the roulette wheel or crap table or shell game. We'd play awhile, till that would get kinda tired, and then we'd lose most of our money. Then we'd go back and dance some more and probably fight the rest of the time, till it was time for us to go to bed.

Most lawmen let the fighters tire themselves out and stayed out of the fracases. When the law stepped into a brawl in Desdemona, the outcome was unexpected.

E. E. Brackens: I drilled once right behind a cafe there when they had a big roughhouse battle one night. There was a deputy sheriff and a Texas Ranger there, and they were trying to stop a fight that had started in the restaurant. One of them hit one of these fellows in the eye with a gun and knocked his eye out with the butt of the gun. That started a roughhouse they couldn't control. The drillers and tool dressers took these two lawmen and loaded them in the car and headed off to De Leon with them. About halfway to De Leon they kicked them out and told them to keep going, not to come back. And they then took everything out of that restaurant and threw it out in the muddy street. I went over there when I got off and everything was out in the mud. And they knocked the partition out between the kitchen and the dining room. And the stove and everything went right out in the mud and in the street. It put the restaurant out of business.

Most bartenders and bouncers avoided brawls, because friends of the fighters often settled up scores later. *Clarence Dunaway* described the revenge of rig builders in Pampa after a young worker was roughed up:

One of their kids was about fifteen, eighteen years old. And he went out there, and maybe he got smart off. But anyhow, [the saloon bouncers] like to beat him dead. He come back, and he was just beat all to the dickens. So they talked to him. And they called old Jim Redman and Jim Neely. Old Neely was a big old Swede. He was about six six or six seven, and he was rougher than a cob. They went and got them a bunch of mattock handles and put them in the car, and they all went out there. They had half a keg of tenpenny nails. And they said, "Just all you

Workers stage a comical photo: no
catch in sight at this pothole in a
Ranger street in 1918.

FISHING AT RANGER.

S-you-know-what clear out. This is closed for thirty days." And they just nailed her up, and they didn't open for thirty days.

Some friends fought every time they met. **V. L. Cox** recalled the meetings of two rig builders in a pool hall:

George Anderson was a big guy, bigger than you are. And him and Bill Wooster would get down in that pool hall, and they're playing. One of 'em would knock the other and put him over the pool table, put him under it. One of the guys who worked down there got more kick out of that. He says, "Them son of a bitches play worse than anybody I ever saw fight." Wouldn't hurt either one of 'em.

Once in a while even veteran fighters had a hard time finding a suitable battle, so they used a smaller friend as bait.

Bobby Weaver: Don and Marshall were big old boys, six foot three or four and weighed two-forty, you know. And they loved to fight. I never will forget when I went to State-Line one time with them. We were working over there north of Lovington. We went down one night to the beer joints and dance hall. I'm little, so when we went in there, they'd get me to go over there and get somebody to pick on me so they could get in a fight with him. And they got me whupped around on real bad. A lot of those guys just loved to fight. They enjoyed it.

Most of the residents of oil camps and towns amused themselves more peaceably, with people-watching and conversation. Two former Wink residents remembered the twenties and thirties.

Alice Keene: We'd hurry and get our evening meal over with and get downtown. All of us would do that, from the other camps, too, you know. We'd all go downtown to watch pedestrians. Well, it was just like going to a movie. One would be dressed so-and-so; and we'd watch them as they'd go into the show and coming out. We'd sit there until nine or nine-thirty, ten o'clock, and watch 'em.

Charles Stroder: Saturday night was the big

Workers and townies hanging out at
the drugstore in Borger in 1927.

night. Everyone would go to town on Saturday night. The men would get in the cars and park. Maybe the men would sit out on the hoods of the cars and talk. But everyone would go to town. The movies were pretty popular. Most families would go to the movies regularly. The majority of the men were pretty sober.

By the time of the First World War, movie theaters had spread across the land. In the new oil towns, they competed with older forms of commercial entertainment, as **Frank Kelly** recalled of Burkburnett in 1921:

> Well, we had some moving picture shows, if you could get into one. They was just like everything else, just crowded and crammed full. And then they had tent shows that would come through here. There's burlesque shows that sometimes wasn't very high class. And then, of course, they had quite a few dances. But brother, if you didn't want to roughhouse, you wanted to stay away from that dance. I never did go to one even when I was a kid growing up out here by myself. I can look back and wonder at a lot of things I didn't get into that I could have.

During the twenties and thirties, major oil companies subsidized semiprofessional baseball teams to provide less risky amusement and to build morale in isolated settlements. As **Dave Brazel** remembered it, playing baseball was a full-time job when he went on the payroll of an oil company:

> They said for us to report to the head roustabout shack, the office, in the morning. So we went up there. The head roustabout started picking his crews and assigning the work. He told this one guy, the foreman we were to work for, "Casey, you take these three ball players." And we looked at each other and kind of giggled, you know, at that. These folks are men, and we're boys. Well, anyway, we went out. Nobody told us anything to do. None of us had ever worked in the oil fields before, and we didn't know what to do. So we were just standing around trying to figure out something to do, and nobody said anything.
>
> The head roustabout went from job to job on his horse. And we saw him coming, so we picked up two-by-fours and tried to walk around look-

ing important, like we were earning our money. He came over to us and took us aside: "Now, look, boys. We hired you guys to play baseball. You mess around those timbers and get a finger mashed or splinters in 'em and lose a ball game, and you're out of a job. We have trouble with these fellas here. When they get off the job, they get home and drink home brew, drink up on that stuff, and get drunk and get in fights with each other and their neighbors and their wives. Since we've had a baseball team we haven't had any of that trouble. And they bet everything they've got on these ball games, so we want you fellas to take care of yourselves."

During the thirties, some companies came to expect more of their ball players.

Clell Reed: I think that was one of the reasons I got hired, because I told the man that I was a baseball player. He said, "Well, what position do you play?" I said, "Anywhere you want me."

A lot of them were ball players. We'd get a little time off once in a while; maybe they'd let us off on Saturday afternoon so we could practice, and Sunday we'd play. In those days all oil companies had baseball teams. And [the players] were either coming down from the big leagues or they were youngsters that's going up. We got no pay; we played for fun. That was our diversion, the only diversion we had.

Now, this started, this baseball, back around Breckenridge and Ranger with Gulf, Magnolia, and all those companies. Then they established the employee relations departments and began to see the fallacy of hiring men just to play baseball. It just faded away, but we did enjoy it in the early thirties.

Some companies continued to support baseball teams through the Depression, but the practice was curtailed during World War II, as *Joe Koesel* explained:

I came out here to play baseball with the Texon Oilers. I used to play professional ball. I just quit and came out here and went to work in the oil field.

Watching pedestrians, a favorite sport
in Wink in 1928.

Every town had a team in those days. Crane, we used to play them all the time. And Mc-Camey had a team. We used to play San Angelo all the time. The first game I played, we went to Wink.

At one time we had fifty-five ball players out here. They'd play a team, and they'd see a good ball player, well, they'd hire him to go to work here so they could play 'em. Yeah, at one time they had fifty-five ball players. All they'd do, they'd send 'em out to cut a weed, something like that. Wouldn't do much. When I came out here, though, we had to work.

It lasted till about the war started. We had to quit. A lot of the boys had to go to the service, and that broke up the club. Oh, they played a little around here. Not too much. They wouldn't hire any more ball players."

After hours, some oil field people looked for more than entertainment. Even before a field was leveled for the ball team, ground was broken for a community church. Workers erected the first buildings, often on land donated by oil companies and other landowners, after their long work days were over. Baptist churches were usually the first separate church buildings to appear in new oil towns, because the first residents came from rural regions in which that denomination was most numerous. In the usual sequence of events, Methodist, Church of Christ, Roman Catholic, and Lutheran churches were built if the numbers of their adherents were sufficient. If there were enough middle-class professionals, Presbyterian and Episcopal missions were organized for them.

Not everyone went to church, however, and the unconverted were not above a joke at the expense of their more orderly brethren, as **William Joseph Philp** recalled of Spindletop:

> Once in a while we would have a big meeting and a very few would come to the Sunday school. One of them boys, I remember, that come to our church . . . he would get drunk. But he finally got converted, and he was one of the roughnecks of that day and time. And there was quite a few of the boys that would come that was good workers and would come to Sunday school and have

More than local celebrities, the Big
Lake Oil Company Oilers pose in
Denver in 1928.

Big Lake Oil Company's OILERS - national semi-pro winners Denver Post tournament - 1928

Top row: Tom Battle, ace fan; W. M. Griffith, bus. mgr.; Chief Harmon, p.; Hi Haven, p.;
Roy Gardner, 1b.; Red Horne, p.; Steve Ellis, c. & mgr.; Ray Johnson, c.;
Jimmie Grant, cf.; Rube Wilhoit, rf.

Bottom row: Horace Wallin, utility, 1b.; Foy Haddock, p.; Gus Leedy, 2b.; Dyke Fuller,
utility, inf.; Trig Housewright, 3b.; Barron McCulloch, lf.; Goldie Rapp, ss.;
Gene Calders, p.

work on Sundays and couldn't always attend. But when we'd have a meeting at night, why, we'd have a good houseful.

We had one old fellow there, he couldn't hardly read. And he'd pick up the Bible, and he read. And he'd get to a certain word, and he'd say, "I can't call that name," and call it "Jim."

One time the boys thought they would have some fun. The Apostolics had a big houseful of people. There was a front door to the little church and a side door at the back end. They thought they'd have some fun, so they get some dog that's running around, and they get tin cans and a stout string, and they fill the cans with rocks and tie to the dog's tail and puts his head in there and hits him a lick. And there he goes right down the aisle. The people all jumped up; they didn't know what was the matter. That was one of the worst little disturbances, and that was more fun than it was doing a criminal act.

Churches were usually built in oil towns and camps after the boomers were joined by production workers, who tended to move less often. In many places, early services were held in homes, tents, or in makeshift buildings.

Mrs. Skeet McAuley recalled the beginning of the Church of Christ in Wickett:

Before we had a church out there, there was a minister used to preach. And he'd come on Tuesday night and have a prayer service. We'd take everything out of the front bedroom and put in a bunch of chairs, and people would come. We had two or three meetings, and then we met down at the schoolhouse.

My great-grandfather was a minister and my uncle was a minister, so I was brought up in the church.

Workers and their families are baptized at Seven Wells, Mitchell County, in the early 1920s.

Services often awaited the arrival of circuit riders or itinerant preachers, who held prayer meetings and baptized in local creeks and ponds. *Pat Rogers* remembered such gatherings during the East Texas oil boom:

We had brush arbor meetings at Joinerville all summer. They'd come down there, and they'd have preaching. They'd have an evangelist-type preacher. Built a brush arbor out of poles and cover the tops with pine limbs and things. They'd have preaching four or five weeks at a time. And while they'd go to the preaching, all the kids were running around and playing hide-and-seek outside. When they'd get all of 'em converted, well, then they'd go to the creek and get 'em baptized.

Company picnics and school events rounded out the social activities of the oil patch, as *Mary Rogers* recalled of East Texas in the thirties:

We'd go to singings and picnics and school [events] just like I'd been through in all my life. I'd take the children, and we just went to picnics and all-day singings and all that kind of thing, you know. We had a friend that had a nice stream running through his pasture. He dammed up part of that and built the nicest swimming pool you ever saw. That was good recreation for the evenings.

An early photograph of an industry
tradition—the company barbecue.

Oh, honey, we had more things to do! Why, every time you turned around, it was a party for somebody. They kept up with all the children, the young folks, you know. The only thing they didn't allow anybody to do there [at White Oak], especially the schoolchildren, seniors and all, was let 'em dance. No, they wouldn't let them dance. So they had to go over to Gladewater. And they had kind of a youth center over there.

A nice place for the children to have their parties and things.

But they had all kinds of school activities and sports. Everybody went wild about football and that kind of thing. And by the time you attended all of those things and went to church and to a picture show and showers and all, why, you just had the calendar full.

Shell employees clown with the handcuffed sheriff at a 1934 company picnic in Wink.

11: White Lightning and Ginger Jake

Work and social life for both businessmen and workers in the oil fields involved liquor. Contractors and suppliers furnished it for their customers. Blue-collar workers drank on return trips from their jobs and during off-hours. Those who came to new oil fields drank in bars set up in tents; those who came to established oil towns drank in grander saloons with mahogany bars and mirrored walls. There was drinking, as well, in city streets, back alleys, and cot houses.

And what an array of liquids went over their palates and down their throats! Before Prohibition, commercial beers and whiskeys were most popular in the oil field. A greater variety of drink—and no less a quantity of it—followed after national Prohibition in 1919. Home brews of all kinds and qualities, fermented in basements, barns, and holes in the ground, enlivened many a gathering. Corn whiskey, the famed "white lightning," stunned, stupefied, and occasionally blinded or killed its eager consumers. More novel intoxicants on the booming market included Jamaica ginger, known more commonly as ginger jake, which produced a burning peppery sensation and, on occasion, serious and permanent damage to the nervous system.

Even during Prohibition, finding strong drink was easy.

Druggists and cafe workers supplied most of the ginger jake, while bootleggers sold home brew, corn liquor, and smuggled foreign whiskey. Much like the organized crime bosses of the nation's larger cities, bootleggers in the oil towns had their own hierarchies and designated territories. They maintained discipline in the ranks with guns and knives. Secure in their standing with an army of local customers, the bootleggers were a law unto themselves.

Local lawmen, sometimes on the payrolls of kingpin bootleggers, often ignored the highly profitable and visible traffic in booze. On rare occasions, usually following a strong protest by the church crowd, as the teetotalers were tagged, the Texas Rangers would swoop down on an oil town and make well-publicized arrests. They usually destroyed some stills, after posing for photographs in front of the evidence, and left the oil town much as they found it. In some instances they gained few arrests and little notice because local lawmen tipped off local criminals of the Rangers' impending arrival.

After the repeal of Prohibition in 1933, workers could again drink beer in saloons and cafes in some parts of Texas. In the dry counties and cities, however, they continued to do business with local bootleggers or to travel to areas that were legally wet. Apart from physical addiction, there was strong cultural pressure to do so: the enduring image of the hardworking and equally hard-drinking and hard-fighting roughneck was durable and compelling, particularly with the strong reinforcement it received from modern advertising. Over the years, the money trade in alcohol in the oil fields has been vast; the cost, measured in broken bodies and shattered lives, is incalculable.

Drinking was an accepted part of oil field life by the time oil was discovered at Spindletop.

Early C. Deane: There was a lot of drinking [at Spindletop and Batson], lot of drinking, lot of fights. But they'd get back on the work next day or two. They'd take Christmas, they called it. I saw one fellow I knew pretty well. I saw him about Saint Valentine's Day and he was drunk. I said, "Why, Bill, my God, you drunk again?"

"Why, no, not again. I'm still taking Christmas."

He turned out to be a mighty fine oil man.

George Ramer: Those fellows would come in out of the field all dirty and grimy in the evening and head for the old Burleson Hotel [in Mc-Camey]. They would get a bath there and clean up some, and then all of them wanted a bit of liquor to wash the dirt out of their throats. It was easy to slip into the habit.

V. L. Cox: Most all the rig builders were heavy drinkers. There was some few that never touched it. Most of 'em drank. Six of us would buy two fifths and drink it before we got back to town. We wouldn't drink on the job, but a lot of us went out to work drunk. Some of 'em got sent home and some of 'em got fired, but most of 'em didn't. The crews all had a favorite spot they went. Most of the rig builders in Odessa used to hang out at Slim Gabrel's pool hall at night. That's where they drank and that's where they'd hire out.

Skeet McAuley: Nearly everybody'd go for a drink after work. Lots of times, if anybody'd goof off or forget, like putting dies on a pipe backwards, something like that, they had to buy drinks for the whole gang. I once carried a set of two-inch dies out there one day to put them on the end of a piece of pipe. I picked one of them up and set it wrong way on the pipe. Everybody started to holler, "You owe us a drink! You owe us a drink!"

Bobby Weaver worked in the oil fields during the sixties:

We always, without exception, put a six-pack in the water can. When we got off work in the afternoon and started to change clothes, everybody got a beer. Sometimes two or three. And a lot of times we'd stop coming in and get some more at a package store. (The crews I worked in really didn't stop often, but a lot of those other guys did.)

Oil field contractors and businessmen learned to adjust their business operations to drinking during the 1920s:

Don Dittman: You found some oil field workers that never touched a drop and you found others that was heavy drinkers. They'd only work for a while. At McCamey, Dad wouldn't pay 'em right off. He'd make an excuse that he didn't have the money to finish the well. (You didn't pay except when you finished a well.) A lot of times we'd make 'em drill two wells before we'd pay 'em off, because we knew they'd be off for a while after they got paid. They'd be off until they ran out of money. Then they'd come back and make you a good hand. Others would drink a little all the time, but they never let it affect them.

Willie Wolf: I wasn't drinking, but I carried whiskey from El Paso and used it in my business. Everybody was drinking. I had people in here who would down a pint of liquor in one shot. This office was a hangout, and they would come in here when they wanted a drink. I carried the liquor from El Paso in drums so they wouldn't catch me. It was Waterfield-Frazier and American, but it was not for sale.

P. O. Sill: Being a rig building contractor, I had to buy a lot of drinks during Prohibition time. I'd get somebody to go into Old Mexico and bring some across to give to those superintendents. I had one superintendent who said, "You can't have the rig job unless you get me a

Young Texon workers wet their
whistles at Fred's Place in nearby
Best, late 1930s.

drink once in a while." And that's the way it was during Prohibition. I had a little fellow whose name was McMurtry, and he says, "I can go over there and bring you back a quart or two." And I'd pay him to go across at Juarez. He would put on a big pair of overalls and a jacket and take a lot of tape. He would tape the bottles to his body and put his clothes over that and bring that liquor across for me so I could give it to those superintendents. We did that quite a bit. I just had to do it.

Oil field saloons came in a wide variety of styles and classes; they also performed banking and welfare functions in most towns. **W. H. Bryant** recalled of Sour Lake and Batson early in this century:

We had some pretty nice saloons, and we had some that was pretty crude. Of course, when saloons first opened up, they generally opened up with two boxes and a one-by-twelve to set whiskey on. Then later they got some pretty good bar fixtures. They run about five hundred to fifteen hundred dollars with mirrors in them. At first you couldn't have a bar with mirrors in it; somebody would shoot them out.

And every saloon that we had was pretty fair to the workin' boys. If one of them got sick, the saloon man was pretty liberal about givin' a loan. We didn't have any hijacking. We didn't have any because they'd get your money quick enough with those crooked dice and cards.

One of the saloons in Batson had a palmetto roof on it, and then in Humble we had one that was made out of slabs from the sawmills, set endways. But that saloon got so bad that if they couldn't get your money, they'd get you drunk and take it off of you in the back rooms. So we got tired of that. One night we got a bailin' line and put it around the saloon and put a clamp on it with a hook. When the log train come by, we hooked it on and scattered that slab saloon about four hundred yards down the railroad track. That's the way we tore it up because it was trimmin' the roughnecks.

Bobby Weaver recalled some Odessa watering holes of the late fifties:

The Odessa bars were on Eighth and Second. There was Danceland at the corner of Second and Dixie. That was a high-class operation. Over on Eighth Street, it was Beer Joint Row. There was the Nip 'n' Sip, the High Hat, the Tivoli, and others. The Nip 'n' Sip was notably tough, and so was the Tivoli. There was Bandanna Lou's on Murphy Street. I never went in Bandanna Lou's; it was pretty rugged.

Friends of mine who were a little older than me, one old boy in particular, talked about going out to Danceland when he was about fifteen or sixteen—you know that age. He said they'd get fights started, boy, just bodacious fights! They like to have wrecked the place one night. He said they had a jukebox in one corner, and he hid behind that sucker. He said there were beer bottles breaking on the wall all around him. It was a real adventure.

Far from discouraging drinking in the oil field, Prohibition encouraged consumption of virtually anything with alcohol content.

Clell Reed: I think they called it Choc beer because it came from the Choctaw Nation in Oklahoma. They made it up there and sold it to the Choctaw Indians. It was made in ten- or fifteen-gallon crocks with malt. It had lots of different ingredients, including added alcohol. Bootleggers made the stuff. They would let it set so long and then they'd bottle it. A bottle of it would have about half an inch of sediment of that malt on the bottom. When you opened it, was what was called wild beer. You had to get on top of it; sometimes it would raise your feet a foot or two off the ground!

White lightning was made in old iron barrels. It would test out at 110 to 120 proof; then they'd thin it down to about 50 to 60 proof. People got killed on it because of the additives. They'd get poisoned on it. You had to know who your bootlegger was and where he was getting his supply. There was a large distillery up near Lovington, New Mexico. It made a specially good whiskey; it sold for a premium. They kept that still well hidden, all covered over.

Red Laughlin: [In Wink] they had these tents, built-up tents, you know. They had this Choc beer they was making. You could go in, get a drink anywhere here. And they was hauling in whiskey from all over. Choc beer cost about a quarter a bottle. Hell, everybody had the damn stuff. Whiskey run about ten dollars a gallon or so. Most of them places they'd sell you a bottle. There wasn't many places where they sold it to you by the drink. You bought the bottle. One old boy had four or five tents back in there. In the middle of the damn floors, he had holes cut out, in the middle of the floors. The Choc beer was buried down in a box. That part of town wasn't nothing but a bunch of damn shacks and tents, some of them with holes in the floor.

R. S. Kennedy: You see, they first made that stuff with a lot of raisins, corn, and a lot of different things. Just laid it up. And it was kind of a mash more than anything. Sour and everything. It'd ferment, settle down, and clear. Then they drank that. They called that Choc. They didn't bottle it, just put it up in gallon jugs. It might near just about knock a fellow out.

Bill Collyns: It was very easy to get a little whiskey from someone living on bootlegging. You knew who they were. You'd approach 'em and go out to the little town of Crossett, west of McCamey a short distance. Or they'd go out in the sand and dig up a half-gallon fruit jar. If you wanted a pint of whiskey, they'd pour you up a pint. If you wanted it colored, that was fifty cents extra. We just took it white at three dollars a pint. The quality was pretty low because it didn't have much age on it, but it served its purpose.

Clyde Barton: Over there in Wink one day, there was this fellow standing across the street and hollering, "Step inside! Fifty-cent overcoat!" I asked this chap I was with, "What in the world was he talking about—a fifty-cent overcoat?"

"Well," he said, "that's a shot of corn whiskey."

They had flophouses over at Wink. This old Ranger captain told me, "Go out behind one of 'em and look." I think those fellows were staying there for twenty-five cents a night. There was a pile of canned-heat cans. There were barrels of 'em out behind tents.

I asked, "How do they get alcohol out of that?"

"Well," he said, "they just take their socks, put it in, and squeeze it out."

Perhaps the most dangerous booze in the oil patch was ginger jake. It was common, and its victims were numerous.

Hood May: I heard that there was stuff, jake, made in St. Louis; had stuff in it that caused jake leg. I drank it myself. I'd buy it at a drugstore. It tasted real hot, see. You pretty nearly had to have a chaser with it. It'd just get your breath. There was a cafe at Penwell. He sold jake that was black. He showed me a jug of it and said, "You want some of that?" And I said, "No! I don't want none of that." It was just black and I knew there was something else in it besides the regular makings of ginger jake.

I think there was some kind of spirits they put in that give 'em jake leg. And I saw girls and men at Wink with it. Their legs would flop, you know. They'd have to tie their feet up so they could walk. One girl, about nineteen years old, had fingers that was bent plumb backwards. She was a nice-looking girl, and she ate at the same boardinghouse I did there at Wink, but she was a pitiful sight. She couldn't hardly hold a cigarette. She had some jake that give her jake leg.

Fred Jennings: Oh, there was a lot of drinking, there was a lot of drinking. They drank this—what they call jake. It's a Joo-maker ginger, but it's got alcohol in it, and they'd take three or four drinks of that, and they'd just climb the wall. Then they'd get out there, and one man'd get soaked up on that jake, and he'd challenge the whole town for a fight.

Burk Paschall: Jake, that was a great drink here [in Breckenridge]. I've seen a lot of 'em that had jake leg. I never knew what jake leg was till this boom came here. But I actually seen it. You can't walk straight, and you have to have crutches and walking canes, and then whenever you step, why, your leg goes to trembling. Your feet goes to trembling. And they're mighty near jerked out from under you. After you get the jake leg, why, you've got it from then on out until it's cured. And there's some of it not curable. They just die with it.

Paul Patterson: Over at Rankin—I used to get a kick out of it—we would hear all kinds of rumors about Crane. They said they had a jake leg race. You know, in those days they would drink Jamaica ginger and sometimes they would lose their equilibrium and coordination. They had a jake leg race. They would get two or three of those old jake addicts and bet on which one would win. They all went in different directions.

Red Laughlin: These guys, skinners and all this stuff, whole bunch of them, they'd come in broke. So damn many. They'd get a payday and they'd all "Hoosier up." Drillers, and tool dressers and everything else, and drink on the goddamn jake. Up there at Breckenridge I saw piles of two-ounce bottles that high in the road, a pile that high. They'd go and buy it by the dozen, you know. Pay two dollars a bottle for two ounces of

that damned jake. Just sit around, pass the bottle around. There were a lot of jake drinkers around here in Wink. There weren't many crippled up, but there was lots of jake around here. But then, hell, that went with everybody. They hauled whiskey in here from all over the country, and one thing and another.

In most towns, bootlegging was controlled by one man, who distributed liquor and designated territories for retail agents. In Wink, the top man was Heavy Brackeen, who moved there in 1927 from Borger. The recollections of several residents delineate his character and attest to his power.

W. F. "Hot Shot" Ash: Heavy Brackeen was the main guy. You'd pay so much: I was paying six dollars a case for beer and fifteen dollars for whiskey. You'd sell the whiskey for a dollar a drink and the beer for fifty cents. Heavy also run the City Cafe.

Floyd L. Carney: He had pool tables in there. Maybe two bartenders and a couple of girls. He had a pretty good business there. I'd say the City Cafe was about seventy-five feet by twenty-five feet. He had an old-timey back bar that was the only fancy thing he had. I think he got that out of El Paso or somewhere. It looked like it was hand carved. Beer was about all we sold. I remember the time I was living at Wink and tending bar for old Heavy. He'd get about four to six hundred dollars and take off. He'd say, "Well, I'm going to be gone." He'd go off to El Paso or somewhere. Like I said, he was a good ol' boy to the last, but I don't think he left Mrs. Brackeen too well off. Those fellas never have any money. Just from day to day. They have a big wad today and tomorrow it's gone. They just made theirs day to day. He'd get a little roll built up and he'd take off and gamble, gamble it all up in El Paso.

J. Conrad Dunagan: I remember Heavy Brackeen well. He was a short, muscular fellow. He was brusque, abrupt, and a little tough. As a beer distributor after Repeal, I never did have any difficulties with Heavy. He was a businessman. But after we got to distributing Budweiser, Heavy's attitude was much improved. He just thought that Budweiser was the finest beer there ever was. He became almost an affable customer.

Lewis Gray: Old Heavy Brackeen was one of them thugs. He was a good man. But he was a thug. Do anything for you, then shoot you in the morning. But he was a nice guy.

Red Laughlin: One of the main guys in here was old Heavy Brackeen. Heavy Brackeen. I'll tell you the same thing I told the goddamn big colonel out of Washington, D.C. They come in here during World War II. They had old Heavy all tied up; they were going to get him for having stolen guns and stuff that belonged to the Air Force, and one thing and another. The colonel and this guy from over there at the Pyote base come into the drugstore I was running then, and they said, "What kind of a guy is this Heavy Brackeen?"

I says, "You want to know about Heavy Brackeen, you should go talk to Heavy. He'll tell you. He's a goddamn racketeer and he don't care who knows it." Heavy didn't care.

Lawmen obligingly pose in front of a
confiscated still in Borger, circa 1929.

F. Ellis Summers: I remember the day that Heavy kidnapped the sheriff. He sent Hot Shot Ash over to the courthouse to make sure I was there. I remember Hot Shot coming in, just as well as if it was yesterday. Came into the courthouse and stood around there a minute. I asked him if there was anything I could do for him. He said that he wanted to know if Mr. Priest was around; said he wanted to borrow his lawn mower. Then Heavy kidnapped the sheriff. Carried him down to his house. There was three or four down there. Hot Shot was there. They told the sheriff that they were going to kill him if he didn't fire me as deputy.

Heavy Brackeen, the last three years he lived in Wink, was as law-abiding a citizen as anybody ever lived over there. He died that way, but he bucked her as long as he could.

Off-duty oil field workers collide in
Snyder in 1950.

12: Sex

In booming oil towns, sex was a highly visible, easily accessible commodity. Prostitutes and pimps arrived in new oil fields hot on the heels of the first roughnecks and roustabouts. Whether in Batson, Breckenridge, or Borger, their approach to plying their trade was usually open and direct; in the here-today-gone-tomorrow atmosphere of a boom, they did not trouble to hide what they did from the respectable folks. Doing business through saloons and dance halls as well as houses, prostitutes were available to young and old "broncs" responding to natural urges. Whether or not they approved, most oil field residents took prostitution for granted.

But that did not mean that prostitutes were accepted on equal terms by other community residents. As elsewhere, the sexual double standard applied in the oil field. While men used prostitutes, people who wanted to be thought respectable shunned contact with them. Though prostitutes ate in boardinghouses, had haircuts at beauty parlors, and visited local doctors, social stigma cut them off from life on Main Street. Of all the uprooted who followed oil field action, prostitutes were probably the most psychologically isolated.

Prostitutes were victims of more than social attitudes. Many had grown up in the rural poverty familiar to their clients. They sought money and excitement only to find a life of hard knocks and violence. Many became alcoholics and drug addicts; some fell victim to the nerve-destroying paralysis that came from drinking contaminated bootleg liquor. Like their clients, they faced the probability of contracting venereal disease. Happiness and fortune did not await them "on the line."

Their lives were ruined by exploitation, but the oil field prostitutes are often remembered for their better qualities, including honesty, generosity, and compassion. Their enduring image is summed up in folklore by Roughneck Pearl, who, frantic to kiss her man good-bye, raced down to the railroad station in Caddo Lake in the nude.

Plummer Barfield, an eyewitness to boom times in Batson, described the girlie shows of eighty years ago:

> The type of women that come in here first, outside of natives, they was just as tough as a boot. They run gambling houses and shows.
>
> They had what they called the Standard Theater. It was strictly for men only. You can imag-

ine what it was. I believe they called the other one the Klondike, and then they had one they called the Crescent. They were operated strictly by that type of woman, with very few men actors in the show. It was pretty much a girlie show, out and out. Language and conduct was just downright vulgar, in modern-day speaking. The jokes they told were smutty and unbecoming to a man, much less a lady. So, that's the type of show it was.

You can imagine the type of girls they was because you couldn't class them as actresses in any modern circuit, because they were just has-beens, drunkards, dope fiends, and what have you. There was some young girls mixed up in there, but most of them were old cats that'd been kicked out of the higher class. Some of them brought young girls with them and made it their business to hustle girls and got a few of them out of the country.

Most new oil towns had red-light districts, where bars, dance halls, and brothels operated under the tolerant eye of local law officers. During the late twenties, Borger's red-light district was infamous.

Earl Snider: Tenth Street, as we know it now, turned out to be dance halls and sportin' houses solid. We had some of the most notorious dance and gambling halls that was ever known. We had the Black Marine, we had the Big Bertha, and we had all kinds of dance halls. One was run by Big and Little Helen. This woman came in here with her daughter and wasn't well-to-do, but she did make her stake and go back to Oklahoma. They also run one on Tenth Street.

[At the dance halls] they called 'em dime dances, but they charged a quarter and danced about forty-five seconds. Some nights they had a special, and they'd have three tickets for a quarter, but the music was just about three times as short. They had rooming houses next door that they'd go to. All the rooming houses sold whiskey, and them that didn't kept girls and sold girls and whiskey together.

Burt Bryan: I was in Mattie's Dance Hall. She had two hundred girls dancing there. Dance

The friendly ladies of the Crosby
House in Batson, 1905.

tickets cost ten cents each. Men would buy thirty or forty at a time. The band—two guitars, a fiddle, and a piano—would start up, and play about half a minute and stop. Then the girls would take another ticket. The girls would take their tickets to a stand and get another kind of ticket worth a nickel. So the girl would get five cents, and Mattie would get five cents. A Syrian fellow I knew danced up thirty-two dollars' worth of tickets in one night.

Gus Keith: If you'd go around to these ten-cent dance halls, you'd see these ladies of the night out there dancing, and they would get these old boys that was about half or three-quarters drunk. These roughnecks, thugs, whatever. And they would get those old boys to buy several dollars' worth of tickets at ten cents a dance. She'd have this old boy give her the tickets, and she'd lead him around. They'd stagger around. Then the music would quit in about half a minute, and she'd tear off a ticket and put it in her sack. And she'd waltz him around and dance some more, and she'd get as many tickets from this old boy as she could. But every once in a while, they would go out back of the building, in the alley, someplace, and they'd come back—both of them would be a little drunker.

The gals, they had these brothels. They had a big plate-glass window like a storefront, and they had a curtain behind them. There'd be two or three of those girls sitting behind there in a rocker or chair, waving and pecking on the window whenever anyone would walk down the street. It was wide open.

C. Don Hughes: You know, on one of these scenes in my old movies, I've got some of those girls—kind of frisky-like—around the rooming houses up there on what was called Whittenburg Avenue. Everybody called it Lysol Avenue. And I used to say when I was showing [the movies], "Oh, the people were so friendly at Borger. If you didn't have a place to stay, these girls would just let you come and stay with them."

Earl Snider: Some of them prostitutes took care of as high as fourteen and fifteen men in the shift of a night. Three dollars. These pimps that these prostitutes keep treat most of them like dogs. I guess that's the way they got 'em into prostitution. These guys dealt in white slavery, in men and women.

The Rangers come to Borger in 1929 and it was quite a spectacle. They come in on the train and that afternoon they run at least three hundred prostitutes out of town. The roads was blocked with 'em.

Many of the pimps and prostitutes moved from Borger to Wink, which had begun to boom in 1927.

Hood May: The whorehouses in Wink were one street over from the main street. They was just old frame houses, and they'd have three or four girls in each one, you know. And they'd have them pimps around there. The pimps didn't do much; seemed like they just took the girls' money, was all they did. People working there, single guys, they'd go down there. And most of the girls were pretty nice. They didn't give you any trouble. One I got to know so well, why, I'd get short of money and borrow off her. One time I asked her: "How come you loan me money?" She said, "Well, I just trust you, that's all." I gambled a little at that time, and when I'd get broke, I'd go to her and get the money. But that's the only one I borrowed off of as far as whores went.

A lot of those girls had been in other oil towns before they came to Wink. I know this one had. She had been in Oklahoma before. She looked like she was part Indian. I don't know; she was some nationality.

One of those girls showed me something once. I went down there and locked my keys in the car. I went back in and told the woman that run the place. So she said, "I'll get into it for you." She got a coat hanger, straightened it out, and run it through the pedal and said, "That's all there is to it." I was about to take the door off or something.

Another time, when I was working at Crane, a guy locked his car up and was looking how to get into it. So I said, "Well, I believe I can get

Dime-a-dance in Borger, 1926.

into it for you, just need a coat hanger." He went over and got one, and we straightened it out, run it straight up that pedal and hooked it down. He said, "I never would have thought of that."

I said, "Neither would I, but a gal showed me how."

Lewis Gray: The first one to have a house there in Wink was Mrs. Webster. And then there was two more, and you'd have a few guys pimping once in a while. I'd say there were eight to twenty guys working. Old Cowboy ran one of them houses. He come from Oklahoma. There were seven whorehouses around, so I imagine there was forty gals. Forty, forty-five, something like that. There were still that many when I came in 1936. Wide open. Man wanted to claim one of 'em, he'd just go down there and say, "Hey, you, right here, two dollars." That's all there was to it.

Former Winkler County sheriff *Ellis Summers*: We had girls on the line. The average was in their twenties. Of course, the people that run those places was old-time madams.

George, what was that gal's name out there, ran that place at Brookfield?

George Mitchell: May Stripp? Wasn't she at Brookfield?

Ellis Summers: [No,] big, heavyset gal. They called it her place. Bessie, Helen? I'd know it if I'd hear it. Ione run that one up there on the north, behind there in Wink. Behind that white building. Shorty Webster was the last one in there, but I think Ione ran it for a few years.

George Mitchell: She moved to California. Dad loaned her three hundred dollars. She said, "I'm giving you a deed and everything to it and if I don't ever come back, it's yours." And she never did come back. She wrote him a letter and told him to do what he wanted to, 'cause he'd just have to take that for the money. We went over there and tore it down and moved it to the ranch, and that's what the ranch house is built out of. We used to remark [to the cowboys] when we was out there, "Boys, if these walls could talk, you wouldn't need TV."

That was the same place that didn't have any screens on the windows. And remember old Odell Hixson, and that other bronc? They saw the justice of the peace go in there. They got 'em a cane fishing pole, put a little old string on it, and tied on a fish hook. The old man pulled off his gun and put it in a chair, pulled off his clothes and got into bed with this old gal. Then those boys stretched that pole and hook over there and got them clothes, dragged 'em back, pilfered the money, and put 'em back.

Ellis Summers: One of them old gals working on the line told me: "Well, I come in on the first load of casing, and I'm going out on the last load of sucker rods."

Unincorporated towns and villages often sprang up near oil towns. Some of them were centers of illegal diversion, much like Grube, near Iraan, which **Gerald "Corky" Huddleston** recalled:

Grube was the risqué settlement near the Yates field. On payday, the girls from Grube and the men who worked them would get into big cars and cruise along the roads. The girls would be dressed in fancy saloon dresses, their legs and garters showing. They'd drive up and down the street. On one side were the workers' wives waiting for some house money and on the other side, the girls from Grube with megaphones calling, "Come on over and see me tonight."

They had an opera house of a kind over there, but they didn't get to use it much because about eighteen months after it was built, the Rangers raided gambling games and blew it up.

One time before we kids were born, my mother told my dad, "I want you to take me over to Grube because I want to see what it looks like." So they drove up one side of the main street and down the other. She told him, "Let's go home; I've seen all I want to see."

Casual sex frequently led to sexual casualties. As both workers and physicians recalled, venereal disease was common.

Earl Snider: I stayed in the bunkhouse with

Sex appeal sells mud in Corpus
Christi in 1943.

the Phillips Petroleum Company one time. There was fifty-five men in there and eighteen of them had venereal disease. On the Mary Huck lease, they had a bunkhouse. One time they moved that bunch out and took them for examinations. There was nine out of fourteen boys that had venereal disease. It was that bad.

In them days, venereal disease was something to be dreaded; treatment was rough, and it lingered on and on. I've seen boys come out and start to urinate; they'd just grab the belt house or the bull wheels or the bull ropes and just cry and scream. And they'd do that sometimes three and four and five times on a shift, twelve-hour shift.

Dr. D. W. Davis: The public in those days [1902] were not well informed. The young men thought it was smart to contract the venereal diseases. They'd brag about it. They didn't know what it meant in the future. They didn't know what was coming later. The majority of cases, in my opinion, the ones I had experience with, they didn't cure.

Pharmacist **Ken Burrows**: First they went to their pharmacist for sweet spirits of niter and if that didn't work they went to the doctor. Sweet spirits of niter is a diuretic. They never wanted it for themselves. A friend needed it. They'd come in and say, "My friend . . ."

Dr. Cecil Robinson: They'd come in and tell you, "Doc, I believe I got the Dog." Of course, that was before the days of antibiotics. It was a lot more difficult to treat. When the sulfa drugs came along, that was a big help. When the penicillin drugs came many years later, I had quit treating acute infections. Earlier we used mercury for syphilis, and arsenicals; and for gonorrhea we used potash and urinary antiseptics.

Prostitutes often were victims of violence, as doctors Cecil and Rose Robinson remembered:

Dr. L. Rose Robinson: The prostitutes were good patients. It would surprise you how many would be stabbed, hurt. Some of it was the prostitutes themselves doing the stabbing, because there were some pretty rough characters moving with oil field activity.

Dr. Cecil Robinson: They suffered some trauma, too. For example, one day a girl from down on Pine Street came walking in and she had her entrails in her hand. She said, "I'd like to have somebody put these back in for me." Somebody sliced her open, you know. We took her in the little operating room we had, gave her some chloroform, sewed her up, and sent her home. She got along fine.

Though social outcasts, when the needs of daily life brought prostitutes into nonprofessional contact with other oil field residents, they behaved decently. Proper behavior off the job often won them a measure of sympathy, but it could never break down the barrier confining them to the underworld.

Dr. Cecil Robinson: They were pretty clean and we made no effort to control their professional lives. They had the same maladies that would strike anybody in the community, and they were good people to treat. They helped one another out. Even Rose treated some of them in

Gladewater [in the 1930s]. Always the highest respect, no ugliness, always helping one another.

Alice Keene: There were a lot of women here, you know, down on the line. Some of these women would come to the beauty shops, but they would never come to the front door of the shop. They'd come in through the back. They would be just as clean-looking and nice as anybody could look. But the operators would tell us to sit on a paper or something on the chair if they'd been sitting in it. We never did sit right down on where they sat.

J. S. Peebles: That was a profession for them. They acted nicer out than other people did. When they was in a crowd, they knew how to act respectable.

Mrs. Sam W. Webb: I can tell you one thing: the women certainly behaved themselves. They were nice, decent girls.

I know there was one young girl. She came in the dance hall down there one night where we were dancing. She wasn't much over sixteen years old. And she saw that crowd of youngsters. She stood there and watched 'em dance for a while. The Holloway girls were there. One of them looked around and seen this girl. Katy Holloway, she started to put on her wraps and leave, and one of the boys went and asked the girl to leave. I've always felt sorry for that girl, because she was just left out of life.

Downtown McCamey, Texas in 1926.

13: Easy Money

Gambling was a common pastime in the South and West, and it flourished in the Texas oil fields. Young men commonly played blackjack, poker, and craps to idle away spare moments between shifts. On paydays, roughnecks and roustabouts, rig builders and tank builders, mule skinners and truckers alike, all sought excitement and fortune by betting their earnings.

It was never hard to find a game. Bars, restaurants, and drugstores had gaming tables in their back rooms. Every hotel in town hosted high-stakes games. Bunkhouses, cot houses, tents, and vacant lots—it did not matter where, workers staked their wages on long shots at instant fortune.

Once in a while, just often enough to sustain the excitement, a roughneck would strike it rich. Much of the time, however, workers were hustled by professional gamblers in local bars and by transient card sharks who posed as workers long enough to rake in fat pots on payday. The professionals ranged from small-time hustlers who moved from camp to camp and field to field, stopping where no one knew their game, to big-time operators who dressed elegantly, ran their own gambling houses, and were local celebrities.

Like so many of the workers they fleeced, the professional gamblers had in common the inability to hold on to money. They, too, fell prey to the thrill of playing for higher stakes with high rollers in bigger games; Heavy Brackeen and others like him lost small fortunes to sharper players in larger towns. When their stakes ran out, they borrowed train fare or gas money and went to the oil field to rebuild their fortunes. Most of these "knights of the green cloth" ended their professional lives as they began them: playing penny-ante games in small-town domino parlors and saloons. In the long run, they gambled and lost.

Bert E. Hull described the action in Sour Lake early in this century:

> After payday we always had a few gamblers that got on the payroll somehow or other. They never looked up anybody's references when they wanted men; why, they just hired them that were around the camp, whoever was there. So, ordinarily, they had three or four professional gamblers on the payroll that they didn't know about. Came payday, why, everybody would start playing cards at night. In two or three days, those gamblers would have every goddamn cent in the camp. In three or four days they wouldn't show up for work.

A crew plays blackjack in Lamesa in
1929, but no money is in view.

They'd take all the money and beat it back wherever they come from, I don't know where.

Gambling was just as popular in the West Texas oil fields of the twenties, according to **Hood May**:

Yeah, there was gambling! Now, in Coleman, there was five or six gambling houses. There was quite a bit everyplace; most oil field towns did have a gambling house. Some of them had more than others. Coleman had more than any oil field town I was around.

In Crane, this fellow had been a gambler all his life. He had a place over there. Nice-looking fellow, always well dressed. He'd gambled all his life, I guess, as far as I know. I never did know where he was from. Didn't know him well enough to ask him. I just knew who he was and what he did for a living.

A guy named Scotty, in McCamey, had gambled all his life. He'd win enough to go overseas and all that kind of stuff, you know. One time some of the baseball players went out to his place in

Best, and he had a crap game. One of the guys had two dollars and fifty cents when he went out there and won eleven thousand dollars. The boys kept trying to get him away from there before he'd lost it, and they finally got him away. They was trying to watch him in town so he wouldn't go back, but he got away from them and lost all of it except what he paid down on a Ford car. Lost the rest of it!

That fellow that had the Tepee Tavern at McCamey had blackjack and dice. I used to go out there. I'd be going to San Angelo for a weekend; I'd sit there, and I'd say, "Well, I need about twenty-five to thirty dollars to spend the weekend at Angelo." And I'd nearly always beat him out of it.

Lots of men gambled their paychecks. [One worker] used to work for the Gulf; he'd lose his check down there at McCamey and just walk out. He was the most unconcerned guy I ever saw. Seemed like it didn't hurt him a bit. Go on about his business till the next payday.

Bill Measures: I never was a gambler. Couldn't see how you could make any money gambling. One evening we was riding the truck and going into town. This old boy said, "Well, I won three dollars." And this one said two dollars, and that one a dollar-fifty, and so on. I counted it up, and they'd won thirty-two dollars, and nobody had lost a cent! Well, when you're gambling, somebody's going to lose. And I couldn't afford to lose any of that fifty cents an hour I was getting laying pipe in 1928.

Hot Shot Ash: I ran booze and gambling in Borger before I came down to Wink. I had gambling in Wink, over on the corner across from where the post office is now. They just called it a gambling house then. It was pretty rough. We had two or three bouncers there; they took care of the fighting and all that. The Rangers raided us, and we got closed up in Wink. They got 123 men and charged 'em twenty-three dollars each. They got all the money on the crap table, on the poker tables, and everything. They probably got a hundred thousand dollars. That was the only time we got raided. The local police couldn't raid us. We had all of them, we controlled all of them. Wasn't no way for them to raid!

Bessie Leonard: My father-in-law always gambled. He did not work. He took what the workingman made. Never had a job. His wife never had a job. Him and another guy was together in gambling. They followed the oil field activity to Best. That's where my father-in-law gambled and had his place, was in Best.

He come and brought me a suitcase one night, and said, "Will you keep this till morning for me?"

I said, "Sure." So I just set it inside of the door and went to sleep, and that was that. I never thought a thing about it.

The next day he come down and said, "Do you know what you kept all night for me?"

And I said, "I have not the slightest idea."

He opened that suitcase, and it was just full of money. Packs and packs of money, rubber bands on them.

After he left Best, he went back to Oklahoma, and he stayed there. He had always been a gambler. None of the children gambled. They were hard workers, you know. Never gambled.

On occasion a professional gambler had the local limelight, as **Mrs. E. W. Purdy** recalled:

He was always into everything, everything. Nobody ever really knew what he did for a living, but they lived well. Well, we all knew what he did: he gambled.

So they called him up for jury duty one day [in 1949]. They were asking him all these questions, and they asked him what he did, what was his profession.

"Nothing."

"What do you do for a living?"

"Nothing."

"What's your occupation, where do you work?"

"Unemployed."

And he went through all this, and finally they said to him, "Well, what did you do when you were last employed?"

Well, it had been a while.

"Well, when and where was it?"

And he said, well, the last time he was gainfully employed, he was employed by Humble Oil and Refining Company as a roustabout in 1932.

He was a professional gambler. He gambled till he died. He was involved in a very big gambling party here in Midland on several occasions. At one time he won and lost the property over there on the corner of Missouri and Wall where the Borden plant is. It was a big ice cream parlor and a big Borden plant, worth an awful lot of money. Well, at one time he won and lost that in a gambling game.

He gambled on his life one time, and the train hit him and killed him. And there was just one train a week.

Gambling was so much a fact of oil field life that many oil patch law officers decided they might as well live with it: some who couldn't lick them joined them. Former Winkler County sheriff **Ellis Summers** remembered:

Those boys that gambled, part of 'em was working boys, part of 'em was pimps, and some were regular professional gamblers in Wink. I guess you'd call 'em professional gamblers 'cause they didn't do any work. They was always around with a poker game. Old Buttermilk was one of 'em, and so was Hardboiled Lewis.

One lady run a place out there at Tulsa, a little camp near Wink [in the early 1930s]. She was one of a big lawman's girls, so they always ran gambling in the back room of that dance hall—

beer joint. Of course, this was during bootlegging days. I done more threatening than I did anything else. I told them, "Now, this is a public place out here, and you're not going to run no gambling joint." So that lasted awhile, but it didn't last long.

I went out there one night when I knew there was gambling in the back room. The lady told me as I came in, "You-know-who told me they could gamble back there." I just kept walking. Of course, there was a big bunch in the dance hall, and somebody ran and told 'em I'd come. By the time I'd got there, they'd scrambled out the window and into the bushes—including that big lawman and another county official.

One of Wink's best-known residents gambled. **George Mitchell** remembered:

Wink was the melting pot of the universe in boom days. They was all there, including Hardboiled Lewis. I don't have any idea what his initials were. He was an old roughneck, would gamble on anything. He lived with the Old Crow till she

Stills and a confiscated roulette wheel are displayed by law enforcement officers, near Teague in the 1920s.

left him. She was in her forties. She kept him and lived with Hardboiled and Did-he-bite-you.

I remember one time I was over at the gambling hall—I never did gamble. Old Hardboiled had just won a big game. So he said, "Old Crow, come here." And he handed her a big bunch of money and said, "I want you to go to Monahans tomorrow and buy you a diamond as big as a horse turd." Then his luck turned. About fifteen minutes after that, he said, "Old Crow, come here. You gonna have to forget about that diamond: I need that money."

Hardboiled had trained old Did-he-bite-you so that he could walk the [bar] rail at the Club Cafe. Old Hardboiled would bet 'em five dollars that the dog could walk that rail. Then that little old dog would get up on that rail and walk down it. Hardboiled would make him stand up, and he'd walk on his hind legs back. For five dollars.

He was a fine dog, Did-he-bite-you was. As part of another bet, he caught the first raccoon in Winkler County, according to Hardboiled.

One day he went over to the hardware store and bought him a lantern and filled that thing full of kerosene and lit it. By that time it was about dark. Old Did-he-bite-you was trotting along there with him. They got up to the Club Cafe, run by old Heavy Brackeen, a rounder and a bootlegger and what have you.

Heavy asked him, "Where you going, man?"

Hardboiled answered, "I was up there at the north end of town to see where they was rigging, and those mesquite trees up there are full of coons."

"Nah, there ain't no coons there," said Heavy.

"They's full of coons," said Hardboiled. "I got old Did-he-bite-you, and I reckon we're going to catch a coon tonight. This old moon'll come out, and we're going to catch a coon. He's the best damned coon dog in Winkler County!"

Well, they bet him many dollars that he would not get a coon. What they didn't know—and old Hardboiled did—was that some local kids had been visiting in East Texas, and they'd brought a little old baby coon back to Wink. Hardboiled saw 'em when he came in off a tour in the north field, and they sold it to him for two dollars.

After the bet was made, anyway, Hardboiled went out and walked over to his room and put the coon in a sack. Then he walked around a back lot a few times. He tossed the sack over his shoulder and walked into the cafe. Old Did-he-bite-you was jumping about four feet to touch that sack. Hardboiled set that lantern up on the bar and said, "Well, I got that coon. Just shook that mesquite tree, and the coon fell out, and old Did-he-bite-you grabbed him and turned him over. Hell, brought him back alive."

Then Hardboiled shook that sack, and that coon hit the counter and then the floor on the run, with the old dog chasing him. Salt and pepper and sugar bowl flew off that counter. Hardboiled gathered up the money, said, "Come on, Did-he-bite-you, we've had a good day!"

Few professional oil field gamblers had much to show for a lifetime of following the booms, as **Benjamin Coyle** reflected:

I'm a fair example of a gambler's life. I'm an old man, seventy-five years old. I have no children. I have no wife. The simple reason: I never would marry because I wouldn't marry a woman that'd have me. Now here I am at the latter part of my life. No children. No nothing. And there's not a businessman in Texas that's worked as hard as I have. I worked the hard hours, too. All of my work's been mostly at night. I've made and lost two or three fortunes.

If I'd have never got gambling in my blood and went into any other business, legitimate business, and worked at it as hard as I have this gambling, I'd perhaps have been Mr. Tom Wyman ["the Gentleman Gambler"] today and had some children or something or other to live for.

And all of my existence, too, that's the crookedest game I ever run.

Texas Rangers guard a chain gang of workers and toughs after a 1926 roundup in Borger.

The Scurry County Jail, circa 1950.

14: A Way to Make a Living

Most people put up with hardships when they left for the oil field. Economic uncertainties, rough living conditions, and uprooting from kin were usual. Many who entered oil field life, however, were young, and the optimism of youth did much to obscure the day-to-day discomfort. As one school teacher, *Gertrude B. Fleming*, said of living conditions in Iraan in the late twenties, "I guess I was just having so much fun I didn't think about it too much." For the young, moving to the oil field meant a chance to get ahead—and for many, taking that chance paid off.

Not surprisingly, those who adapted successfully to the terms of life and work, those who found the economic advancement they hoped for, see the brighter side of oil field life. They admit that it meant putting up with hardships, but they usually hurry to point out that benefits have outweighed hardships. At the very least, they argue, the oil field offered them a better life than they would have enjoyed back on the farm. Though they had to leave kin behind, they are quick to point out that they made many new friends. If they once lived in tents and shacks, they proudly show the comfortable homes they now enjoy. With good reason, they are proud of their oil patch identity. For them, life in the shadow of the derrick has been rewarding.

It is more difficult to find those who take a less positive view. Those who never liked the oil field but still live there are usually reluctant to express their discontent. More to the point, those who could not bear living in the oil patch usually did not remain there; they got out and stayed out.

By the late 1950s, the great exodus of rural Texans to the oil fields was over, for the most part. As it became harder for small farmers to endure, their ranks thinned and there were fewer farm boys: the reserve of ambition and muscle that developed so many oil fields was depleted. At the same time, oil field work—hard, dirty, and dangerous—lost appeal when young workers found easier jobs that paid almost as well, in Houston, Dallas, Fort Worth, and the other growing cities of Texas. The search for economic opportunity took new directions, away from the oil patch towns. The nomadic days of following booms and living out of a suitcase or two were past. Workers bought mobile homes and tract houses with mortgage money; their children stayed in school. By the 1970s, life in the oil patch was more settled and less rugged than it once had been. The traditional oil field lifestyle was largely a thing of the past.

The oil field attitudes and values that still survive are largely those that early workers brought from the farm: tol-

eration of work that is sometimes hard, dirty, and dangerous; acceptance of the fact that jobs and fellow workers come
and go; agreement that grit counts for more than cleverness
and that football is more important than calculus; belief that
in the end it is important to take root in their own pieces of
Texas, as their fathers and grandfathers had hoped to do, to
stop the incessant movement from place to place that *Vera
Lacefield* recalled:

After you're in it so long, it's just routine. You
get used to it. We'd have our two weeks' vaca-
tion in the summertime; well, we'd take it in the
summertime. We looked forward to our vaca-
tion. Come to Oklahoma, that's where John's
folks lived.

Part of it I wouldn't mind doing over. Some,
I'd just rather forget it. You know, just like everything else. It has its good points and bad points.
In some places we had nice places to live, and
some of them wasn't. But we'd know we was
going to move on.

Anne Swendig: When John first graduated
from college and we moved five times the first

year in his training program, and we had three
kids, my relatives were absolutely appalled.
They said, "Well, Anne, you can't do that. I
mean, you have children, you can't move them
out. I mean, how are they ever going to get any
education? You can't do that. Why, they're just
going to be ruined, you know."

And I said, "I don't see what you're worried
about. I mean, we're only moving, we're probably only going to move once or twice a year. And
the kids' daddy moved, lived in forty-three different towns and went to twenty-eight different
schools. He managed to graduate from college;
he's got a darned good job. Why should I worry
about these kids if they have to move probably
half as much as what he did?" Nowhere near as
much as half, as it turned out. There's a lot of
advantage for kids to be able to adjust. They
learn an awful lot by adjusting that they wouldn't
have learned staying in one place.

Mrs. Clell Reed: It was a very interesting life
because you moved. At that time [the thirties
and forties] the people moved quite often. But

they don't do that anymore, not as much, I guess. I still say people who have lived all of their life in one spot have missed an awful lot, because of the meeting of different people.

No, I don't have any regrets, because I think everything that we have gone through and everything that we have learned and the wonderful friends that we have made over these years would cause anyone not to have any regrets.

We've had a very interesting life. We've had a lot of hard times. But everybody else was having them the same as we were, and you don't know that people are having hard times when it's like that. Everybody was in the same boat.

Bill Briggs: I traveled an extensive area, really. Midland kind of was a hub, [but I] did not take an apartment in Midland until 1951. I'd be in Snyder for six weeks, and Odessa for nine weeks, and Monahans for a while, and Crane for a while, and McCamey for a while. To me, the one thing that always titillated my fancy was picking up the Fort Worth newspaper and opening up to the oil section, which, of course, gave me a hint as to where I was going next.

There were always interesting people. And you would run into a lot of the same people. Not credit people such as myself, but they were people who were there with their own businesses. One particular man that I've been friends with for years was a communications man with the Atlantic Pipeline Company. I would run into him five or six times a year. You formed a kind of cadre.

It was a rather wild and hectic life. And many of them were rather wild and hectic people—and didn't survive it.

Many who have lived in the oil field mention the economic attractions of oil field work, particularly with respect to weathering hard times and to getting ahead. *Mrs. Joe Koesel* lived most of her adult life in Texon:

It didn't bother me the way it did a lot of other people. A lot of women just went berserk, what with the dirt and the dust and everything. But why let anything do you that way when if you didn't like it, you could get up and leave? And I didn't want to leave.

During the Depression we made $117 a month. Both his parents were living; we kept his parents. We bought a new car. We took trips. We didn't know we had a depression out here. You'd get a hundred miles away, and you saw people that were going hungry.

And we lived real good. We didn't have luxuries, but we had what we needed. Everybody [in Texon] was helping somebody, other people off from here, where the Depression hit so badly.

Joe Koesel: Yeah, it was pretty rough in other places.

Mrs. Koesel: This was the best place during the Depression to work.

Bill Ingram spent many years working for Gulf:

When it came to book learning, I didn't have it. I didn't have a chance in the world to go to a school. Fifth grade is as high as I ever got in school, but they ain't nary another fifth-grade man in Texas or Louisiana ever got as high in the company as I did. Came here [Breckenridge] in 1919 and still here. Still drawing a check from the Gulf. The Gulf's been wonderful to me. 'Course I put out a lot of work, but way back in those early days, there was a lot of things come up, lot of things.

If I was gonna start out again, I'd go to work for an oil company. Because if a man'll take an interest in it, it's interesting. And it's good pay, and you just got more privileges. You got a better life. Look, Lord-a-mercy, here I am. I've put in my hard day's work. I get enough now to sit down and do nothing on, see. I got a good living coming. State don't pay me nothing. I don't have to depend on a lot of things that other people do. I've made lots of money, I've spent lots of money. Lot of people do. . . .

No regrets to my life at all that way. 'Course, I probably lived a sinfuller life than I should. I never was in no meanness or nothing, but I didn't live as good a Christian life as a man should do.

Life goes on: home-packed lunch buckets and cafe-filled brown bags are emptied by rig builders near Borger in the 1940s.

Clell Reed worked for Humble:

Nineteen-seventy-two, December the first, I retired. All in all, I had forty years' service with Humble, which is now Exxon. And I wound up my Humble career at Odessa, Texas, as a material coordinator.

I had a good job, I thought, because I had many, many different jobs, and I had come up the so-called ladder one step at a time. I tried very hard not to miss anything that the company had to offer in the way of an educational process or school. I never missed one. I stayed in the library a lot at night because I had to, because I was a high school dropout. I tried very hard to keep up. I was just determined that they [fellow workers] wasn't going to pass me, and they were nipping at my heels all the time.

Personal satisfaction from work, as well as economic benefits, appealed to many workers in oil field jobs. *Bobby Weaver* worked as a tank builder in the fifties:

It's very specialized, very, very specialized.

And you got the attitude—the general attitude was "Nobody can work as hard as we can." That's where the pride lay. And there's something else about tank building. I still think this way, you know; it's kind of like writing books. Once you go back that way, you can always point out there and say, "I built those." They're there. You have a monument to your work. Drill an oil well, there ain't much to it.

George C. Webber was a geologist:

I stayed in for the challenge, the challenge to unlock secrets that the earth has hidden deep within. The challenge to try to be a little smarter than some of the others before me and after me. The challenge to find oil. You suffer a lot of heartaches and die by inches when you're watching that bit go deeper and deeper, and you're supposed to be cutting into your oil sand, but you don't. But when you do find the oil, the sense of accomplishment is tremendous, something I can't explain. I'm proud to have been a small part of the history of the oil industry in this great state, and that's what makes the past twenty-three years all worthwhile.

Oil field life had its unpleasant side, as **O. G. Lawson** remembered:

> Conditions were very much the same [from place to place]. Sometimes I look back over them, and they seem now as if they were almost impossible. Kind of a bad dream. The dirty boarding-houses, the very poor beds, the extreme hardships we worked under, and sometimes gas wells that they had no way of controlling—people would burn—and the crude way of working. Sometimes it all seems now as it couldn't have been. Cannot imagine how people could have worked and accomplished what they did under those circumstances, worked under those conditions. The awful muck, the work we did, the teams . . . moving some of the larger boilers, . . . the men all swearing and arguing what to do.

> Certainly I don't want to go back to the good old days! I am perfectly satisfied the way I am now. I often [wish] that I would be a boy now and just starting all over: have some sense the next time coming up.

J. R. Miller saw fortunes made and fortunes lost:

> I never will forget during the awful drouth of 1917 and '18 all over West Texas. I was in business in Ranger, and a man from Stanton, Texas — he was in the hardware business there, and the drouth got him and he had to close out—he drove to Ranger in an old run-down Dodge car. He came to me and wanted me to sell it for him if I could. I sold that old Dodge car for him. He took that money. In Ranger there was a big boom. Oil was over four dollars a barrel at the casinghead. Proration was unheard of. Leases just kept advancing. I don't care what you paid for them, they would just go higher. . . . This old boy I sold the Dodge for, the first thing you know he had a barrel of money on hand. He kept buying leases, and they kept taking them away from him, you know, with oil four dollars a barrel, the World War on, and begging for oil. When Mexia came in, he went to Mexia and drilled his own well and struck a dry hole. Do

you know where I saw him in the next ten years? He was collecting for Sears and Roebuck in Fort Worth, Texas.

I observed this during the big boom: there's very few people can take it with money. It will ruin them. I had friends that I thought was solid as a rock in Ranger, Texas, and when they hit it rich, the first thing you know the home was broken and gone. Other men will go right down the line.

I'll tell you this about money: there's very few people that can take it. It just goes to their heads. They just fly high, drink, and chase women.

I will say this about an oilman: I believe he is the best loser on the face of the earth. He will go out and spend his last dime on earth that he's got, and he'll walk off from that dry hole and never look back, saying, "Well, I know where I got that start; I'll go back and get it again."

The ups and downs of oil field life were often hard on children, some of whom vowed not to follow in their parents' footsteps. A driller's daughter reflected on her family's experience:

This is what took my dad into the oil field: to get away from farming—which may be kind of usual, I don't know. He ran away from home when he was about fourteen, I think. He hated the farm. I think that when he was growing up, maybe those were the years when farmers were just really having some really difficult times.

He was—I'm sure it's not an unusual story— he was really going to be the one who got rich off the oil field. He thought he had it made for a period there. From very early on, he was able to move up and form a little company with a man, and they did drilling, contracting. They had their own company. They, of course, had the usual falling-out. My dad was taken to court and lost. We don't know if he actually declared bankruptcy or what, but he just really hit the bottom. This was the Depression period.

Part of my negative feelings [about oil field life] are that he was away from home so much. He went to the East Texas oil field; that would have been about '32, something like that. He went there to work. And we were four children left with my mother. She became the hero for us, because she was very resourceful. How she managed, I don't know.

My mother never allowed us to feel critical of him, for the work he was in. But yet, at the same time, she didn't like it either. She got across this idea, "Well, those are just the oil field people." But, see, that wasn't us. "Well, they're just oil field people."

I remember starting in to school, when we asked, "Well, what are we going to say when the teacher asks what your father does?" It was never just to say that he was a driller, but you'd say that he's a rotary *contractor* and driller. And that, I guess, gave her status—although she was not a person to put on airs at all. But it did mean something to her.

One of my high school friends was from one of the families that became immensely wealthy off of the oil field. They were ranchers originally. She married one of the oil field men, one of the young guys who came in.

So here I went off, and I worked for twenty-five or fifty cents an hour or something to go through college and do well by myself. When I went to my dad's funeral, in came this woman who looked like somebody who had spent her life on a ranch. It was that girl. Of course, we felt a real affection for each other. But the first thing that struck me was "Look how I have worked my tail off to get out of this [the oil field], and you are immensely wealthy, and you have spent your life here." Isn't that strange? I mean, it's not strange at all, but, my gosh, if I had had all that money, which they have had over and over, wouldn't I be living on the California coast!

All I ever thought about [the oil field] was getting out of it. And I wonder, because I did have these friends. But I sensed that it really was the pits. And I was right. I mean, that place [the oil town in which she grew up], who would ever want to live there? My idea was just to get away.

I'm sure that I'm just basically a snob. I'm a Texas oil field girl who's a snob. I think that's it. I was quite a student. I was serious. I wanted to do something different.

I think I told you that my son has gone into this work. He's an oil field driller now. Oh, that's my comeuppance, you know. Oh, sure. Because I made it to the East Coast, and I brought my kid up with the museums and the Tiny Tots Symphony. He made this trip out to New Mexico. Saved his paper route money. And he fell in love with the West. Then, when he fell in love in college, he and his wife, they picked up and went to New Mexico. He didn't know about the oil field then. They didn't care how they lived, they just loved that existence out there. Then he began to see that he had to get work, and he kind of fell into the oil field work. And he's Grandad all over.

Of course, I really gave him a hard time. So my last plea to him—see, I guess this is really the key for you to my snobbery—my last plea was "Well, what about the people you're going to be associated with?" And my son said, "Well, Mother, they're so colorful! They're so colorful!" Then, the next day he was telling me about one

of his associates who had had a seven-year sentence in San Quentin or some place like that. He slipped and told me. I said, "Now, there's your colorful personality! Seven years is what you get for killing people."

But he still has stuck with it. Which has been a bitter pill for me. He's a driller. He thinks it's great.

Patience Blakeney Zellmer recalled that her parents wanted more for their children than they thought the oil field could offer:

My daddy would always drill in our heads, "Do not marry a roughneck, do not marry a roughneck!" Well, some of them had not graduated from high school; most of them had not ever had a formal education in college. That was taught me: "You aspire for higher things." Mother and Daddy would tell us, "If you're going to have a better life, you're going to have to marry a boy that will give you a better life."

Daddy would not allow his boys to work in

the oil field. They did not want their boys to get hurt. See, it was very dangerous to work out there on the rig. They were so afraid of their boys getting hurt.

Daddy and Mother were very concerned about our welfare, getting out of Wink into a better life. They did not want their children just to see that. I was raised by a man that had high ideals and standards, and he wanted something much, much better for his children. So we were motivated by that.

For all that many oil field residents admit that the oil field had drawbacks, most of them have come to identify with the interests of the petroleum industry. They often react strongly to criticism of the oil industry by journalists and politicians. They understand that legislation unfavorable to the industry is likely to mean fewer jobs and more neighbors out of work. Above all, they believe that Americans outside the oil field do not understand the importance of the industry that has shaped their lives.

Clell Reed: All the people, like up in the Northeast, Boston area—not to cite person-alities, you know what I'm talking about—that's raised so much cain about the big oil companies, they never refer to them as oil companies, it's always the big oil companies. They've picked on the oil industry that's paved more roads, built more schools, educated more children, and have done more for the United States than any industry alive today. I dare anybody to say that's not true. And it sure does make you feel bad, for them to keep imposing this and imposing that. They make them look like monsters. They always bring in the "guilty" before you're proven innocent. It makes us feel very badly about that, and we're very particular about who we vote for in politics.

Tom Wilmeth: Somewhere in your book, if there's any way you could mention it, I'd love for you to tell everybody that we're all in the oil business. Everyone on the face of this earth is at this time. Someday your baby is going to see something different, but that's a long ways off. Sometime, if you can, sit down and name me ten things that aren't petroleum related. One thing I can think of that has nothing to do with oil, and that's the sun. But what I'm speaking of is material things. Everything in this room is petroleum related: it took petroleum to bring it to us for us to enjoy. People just don't realize.

If they had it to do over again, would they? Most oil field residents who talk about the past would agree with retired oil field supply salesman *Carl Angstadt*:

Well, it's been an interesting forty years. If I had my life to live over again, I'd like to. I don't believe I'd change it. I believe I'd like to go through it again. I've worked pretty hard. The oil country, I think, is an interesting life. There's changes every day.

I've had a lot of experiences. I've had a lot of hard life in the way of living, living conditions.

But I wouldn't trade it for anything that I know of.

Senator Lyndon B. Johnson seeks the support of oilmen and ranchers at the dedication of Snyder's new country club in the early 1950s.

Interviewees

The people listed below are quoted in this volume. Following the interview data, an abbreviation identifies the public depositories of the interviews that were not conducted by the authors.

PBPM = The Abell-Hanger Collection, Permian Basin Petroleum Museum, Library and Hall of Fame, Midland, Texas.
PPM = The Archives of the Panhandle-Plains Museum, West Texas State University, Canyon, Texas.
PTO = Pioneers of Texas Oil Collection, Barker Texas History Center, University of Texas at Austin.

Cullen Akins, interviewed by S. D. Myres, June 17, 1970, Odessa, Texas, PBPM.

Mr. and Mrs. W. A. Allman, interviewed by S. D. Myres, June 19, 1970, Crane, Texas, PBPM.

Carl Angstadt, interviewed by Mody C. Boatright, August 4, 1952, Eastland, Texas, PTO.

Anonymous, interviewed by Diana Davids Olien, February 8, 1985, Midland, Texas.

William F. "Hot Shot" Ash, interviewed by Roger M. Olien, March 2, 1979, Hobbs, New Mexico.

Plummer Barfield, interviewed by William A. Owens, August 1, 1952, Sour Lake, Texas, PTO.

Clyde Barton, interviewed by Roger M. Olien, April 6, 1979, Kermit, Texas.

Bill Beckham, interviewed by Roger M. Olien, May 8, 1984, Kermit, Texas.

Mrs. John Berry, interviewed by Diana Davids Olien, April 6, 1984, San Angelo, Texas.

W. H. Bond, interviewed by Roger M. Olien, October 12, 1979, Midland, Texas.

Hugh Boren, interviewed by Roger M. Olien, February 21, 1978, Snyder, Texas.

Mrs. Robert Boykin, interviewed by Diana Davids Olien, January 27, 1978, Midland, Texas.

E. E. Brackens, interviewed by S. D. Myres, July 16, 1970, Wink, Texas, PBPM.

Dave Brazel, interviewed by Roger M. Olien, July 28, 1980, Big Spring, Texas.

Bill Briggs, interviewed by Roger M. and Diana Davids Olien, August 20, 1977, Midland, Texas.

Burt Bryan, interviewed by S. D. Thompson, April 17, 1973, Borger, Texas, PPM.

W. H. "Bill" Bryant, interviewed by William A. Owens, July 29, 1952, Sour Lake, Texas, PTO.

Ken Burrows, interviewed by Roger M. Olien, April 6, 1979, Kermit, Texas.

Floyd L. Carney, with F. Ellis Summers and George Mitchell, interviewed by Roger M. Olien, May 22, 1979, Kermit, Texas.

Walter Cline, interviewed by Mody C. Boatright with Louise Kelly and J. W. Williams, August 13, 1952, Wichita Falls, Texas, PTO.

William H. "Bill" Collyns, interviewed by Roger M. Olien, May 9, 1979, Midland, Texas.

V. L. Cox, interviewed by Roger M. Olien, April 25, 1980, Odessa, Texas.

Benjamin Coyle, interviewed by William A. Owens, July 30, 1953, Houston, Texas, PTO.

Mr. and Mrs. C. E. "Steve" Cullum, interviewed by Roger M. Olien, August 12, 1980, Wickett, Texas.

C. J. Davidson, interviewed by S. D. Myres with Berte Haigh, March 19, 1971, Forth Worth, Texas, PBPM.

Dr. D. W. Davis, interviewed by William A. Owens, July 18, 1953, Beaumont, Texas, PTO.

Paul Davis, interviewed by Roger M. Olien and J. Conrad Dunagan, June 24, 1982, Midland, Texas.

Early C. Deane, interviewed by William A. Owens, July 10, 1953, Houston, Texas, PTO.

Don Dittman, interviewed by Roger M. Olien, April 6, 1979, Kermit, Texas.

J. Conrad Dunagan, interviewed by Roger M. Olien, April 19, 1978, Monahans, Texas.

Clarence Dunaway, interviewed by Roger M. Olien, April 30, 1980, Odessa, Texas.

Mrs. I. L. Edwards, interviewed by Diana Davids Olien, May 30, 1979, Midland, Texas.

Gertrude B. Fleming, interviewed by Diana Davids Olien, February 12, 1985, Midland, Texas.

W. H. "Steamboat" Fulton, interviewed by Roger M. Olien, October 27, 1979, Midland, Texas.

Ruth Godwin, interviewed by Roger M. Olien, March 16, 1978, Kermit, Texas.

Helena Grant, interviewed by Diana Davids Olien, November 14, 1983, Midland, Texas.

Lewis Gray, interviewed by Roger M. Olien, May 22, 1979, Kermit, Texas.

Frank Hamilton, interviewed by Mody C. Boatright, July 29, 1952, no place given, PTO.

H. A. Hedberg, interviewed by S. D. Myres, June 26, 1970, Fort Worth, Texas, PBPM.

Karolyn Hendrix, with Mrs. Clifford Lyle and Tom Wilmeth, interviewed by Diana Davids Olien, October

26, 1978, Midland, Texas.

W. Horace Hickox, interviewed by Don Abbe, March 15, 1982, Borger, Texas, PPM.

Virginia Hoffman, interviewed by Diana Davids Olien, February 17, 1984, Midland, Texas.

Robert Horn, interviewed by Betty Orbeck, June 16, 1970, Odessa, Texas, PBPM.

Mr. and Mrs. J. M. Horner, interviewed by S. D. Myres, July 13, 1970, Wink, Texas, PBPM.

Gerald "Corky" Huddleston, interviewed by Roger M. and Diana Davids Olien, August 10, 1982, Midland, Texas.

C. Don Hughes, interviewed by Bobby Weaver, July 23, 1982, Amarillo, Texas, PPM.

Bert E. Hull, interviewed by William A. Owens, August 24, 1953, Dallas, Texas, PTO.

Bill Ingram, interviewed by Mody C. Boatright, July 29, 1952, Breckenridge, Texas, PTO.

Fred Jennings, interviewed by William A. Owens, June 19, 1952, Goose Creek, Texas, PTO.

Dr. Homer Johnson, interviewed by Diana Davids Olien, January 30, 1978, Midland, Texas.

Alice Keene, interviewed by Diana Davids Olien, March 21, 1978, Wink, Texas.

Gus Keith, interviewed by Bobby Weaver, March 4, 1982, Amarillo, Texas, PPM.

Frank Kelly, interviewed by Tex Yeager, February 25, 1978, Burkburnett, Texas, PPM.

R. S. Kennedy, interviewed by Mody C. Boatright, July 30, 1952, no place given, PTO.

Mr. and Mrs. Joe Koesel, interviewed by Diana Davids Olien, March 27, 1982, Texon, Texas.

Vera Lacefield, interviewed by Diana Davids Olien, January 3, 1981, Midland, Texas.

C. P. "Red" Laughlin, interviewed by Roger M. Olien, April 6, 1979, Wink, Texas.

O. G. Lawson, interviewed by Mody C. Boatright, July 28, 1952, Cisco, Texas, PTO.

Bessie Leonard, interviewed by Diana Davids Olien, May 19, 1978, Midland, Texas.

Loreet Loftin, interviewed by Roger M. and Diana Davids Olien, March 27, 1982, Big Lake, Texas.

Mrs. Clifford Lyle (Martha), with Karolyn Hendrix and Tom Wilmeth, interviewed by Diana Davids Olien, October 26, 1978, Midland, Texas.

W. E. Mapp, interviewed by Roger M. Olien, April 23, 1980, Odessa, Texas.

Hood May, with Tom H. Neel, interviewed by Roger M. Olien and J. Conrad Dunagan, April 19, 1978, Monahans, Texas.

G. C. "Skeet" McAuley, interviewed by Roger M. Olien and J. Conrad Dunagan, March 24, 1984, Monahans, Texas.

Mrs. G. C. McAuley, interviewed by Diana Davids Olien, March 24, 1984, Monahans, Texas.

William H. "Bill" Measures, interviewed by Roger M. Olien, February 16, 1979, Midland, Texas.

Mr. and Mrs. R. V. Melton, interviewed by S. D. Myres, June 10, 1970, Crane, Texas, PBPM.

J. R. Miller, interviewed by S. D. Myres, February 9, 1970, Midland, Texas, PBPM.

George Mitchell, with Clyde Barton, interviewed by Roger M. Olien, April 12, 1983, Kermit, Texas.

George Mitchell, with F. Ellis Summers and Floyd L. Carney, interviewed by Roger M. Olien, May 22, 1979, Kermit, Texas.

W. C. "Bing" Moddox, interviewed by Don Abbe, February 9, 1982, Borger, Texas, PPM.

Orville "Checkbook" Myers, with Charles W. Chancellor, interviewed by S. D. Myres, October 29, 1971, Midland, Texas, PBPM.

Harry R. Paramore, interviewed by William A. Owens, July 2, 1952, Beaumont, Texas, PTO.

Burk Paschall, interviewed by Mody C. Boatright, July 30, 1952, Breckenridge, Texas, PTO.

Paul Patterson, interviewed by S. D. Myres, June 18, 1970, Crane, Texas, PBPM.

J. S. Peebles, interviewed by Roger M. Olien, May 15, 1978, Wink, Texas.

William Joseph Philp, interviewed by William A. Owens, July 17, 1953, South Park, Texas, PTO.

Mary M. Porter, interviewed by Diana Davids Olien, December 11, 1978, Midland, Texas.

O. C. Profitt, interviewed by Roger M. Olien, October 20, 1978, Odessa, Texas.

C. O. Puckett, interviewed by Roger M. Olien, February 4, 1978, Midland, Texas.

Mr. and Mrs. E. W. Purdy, interviewed by Diana Davids Olien, April 10, 1984, Midland, Texas.

George Ramer, interviewed by S. D. Myres, February 12, 1970, McCamey, Texas, PBPM.

Frank Redman, interviewed by William A. Owens, July 20, 1953, Beaumont, Texas, PTO.

Clell Reed, interviewed by Roger M. Olien, April 6, 1984, San Angelo, Texas.

Mrs. Clell Reed, interviewed by Diana Davids Olien, April 6, 1984, San Angelo, Texas.

Dr. Cecil Robinson, interviewed by Roger M. Olien, April 6, 1979, Kermit, Texas.

Dr. L. Rose Robinson, interviewed by Diana Davids Olien, May 22, 1979, Kermit, Texas.

Mary Rogers and Pat Rogers, interviewed by Diana Davids Olien, May 21, 1980, Midland, Texas.

John Rust, interviewed by Mody C. Boatright, September 12, 1952, Borger, Texas, PTO.

Allie V. Scott, interviewed by Diana Davids Olien, March 23, 1978, McCamey, Texas.

Mr. and Mrs. W. W. Silk, interviewed by Mody C. Boatright and Louise Kelly, August 15, 1952, Wichita Falls, Texas, PTO.

P. O. Sill, interviewed by S. D. Myres, May 12, 1970, Midland, Texas, PBPM.

H. P. Slagel, interviewed by S. D. Myres, December 5, 1969, Colorado City, Texas, PBPM.

Earl Snider, interviewed by Mody C. Boatright, June 19, 1952, Borger, Texas, PTO.

Mrs. Joe Starkey (Theola), interviewed by Diana Davids Olien, August 14, 1980, Monahans, Texas.

Charles Stroder, interviewed by Roger M. and Diana Davids Olien, August 21, 1980, Crane, Texas.

F. Ellis Summers, with Floyd L. Carney and George Mitchell, interviewed by Roger M. Olien, May 22, 1979, Kermit, Texas.

Anne Swendig, interviewed by Diana Davids Olien, May 31, 1979, Midland, Texas.

John Swendig, interviewed by Roger M. Olien, May 28, 1979, Midland, Texas.

Mr. and Mrs. Hoke Tehee, interviewed by Roger M. Olien and J. Conrad Dunagan, October 11, 1979, Monahans, Texas.

Bobby Weaver, interviewed by Roger M. and Diana Davids Olien, July 10, 1984, Canyon, Texas.

Mrs. Sam W. Webb, interviewed by William A. Owens, September, 1952, Fort Worth, Texas, PTO.

George C. Webber, interviewed by Minda Webber, April 9, 1977, Abilene, Texas, PPM.

Tony Wilburn, interviewed by Roger M. Olien, April 6, 1979, Kermit, Texas.

Mrs. James Williams, Sr., interviewed by S. D. Myres with George T. Abell, September 15, 1971, San Angelo, Texas, PBPM.

Tom Wilmeth, with Mrs. Clifford Lyle and Karolyn Hendrix, interviewed by Diana Davids Olien, October 26, 1978, Midland, Texas.

Ben Wilson, interviewed by Roger M. Olien, February 23, 1978, Snyder, Texas.

Mr. and Mrs. R. V. Wilson, interviewed by S. D. Myres, June 16, 1970, Odessa, Texas, PBPM.

L. E. Windham, interviewed by S. D. Myres, February 11, 1970, Rankin, Texas, PBPM.

Mr. and Mrs. William "Willie" Wolf, interviewed by S. D. Myres, February 12, 1970, McCamey, Texas, PBPM.

Mrs. Julius Zellmer (Patience Blakeney), interviewed by Diana Davids Olien, September 9, 1983, Midland, Texas.

Photo Credits

The photographs in this book were graciously provided by the following:

Barker Texas History Center, University of Texas at Austin, Pioneers of Texas Oil Collection: pages 3, 19, 41, 57, 195, 215, 233

East Texas Oil Museum, Kilgore College, Kilgore, Texas: pages 17, 24, 47, 49, 51, 151, 170, 179

Farm Security Administration Collection, Library of Congress, Washington, D.C.: pages 60, 62, 67, 77, 86, 100, 125, 131, 132, 133, 221

Mrs. Gertrude B. Fleming: page 197

Mr. and Mrs. G. C. McAuley: pages 102, 157

Roger M. and Diana Davids Olien: page 96

Archives of the Panhandle-Plains Museum, West Texas State University, Canyon, Texas: pages 106, 145, 186, 209, 218, 236

Permian Basin Petroleum Museum, Library, and Hall of Fame, Midland, Texas:
Abell-Hanger Foundation Collection: pages 11, 34, 53, 91
J. D. (Boss) Bonner Collection: page 45
Buckley Collection: page 228
Lilla Beyer Carter Collection: pages 116, 121
Willard Classen Collection: page 193
Delz Collection: page 202
James (Blondie) Flowers Collection: page 36

Frank Forsythe Collection: page 113

Roy F. Gardner Collection: page 191

Gilcrease Oil Company of Texas—E. L. Ames Collection: pages 14, 75

Humble Pipeline Company Collection: pages 108, 118, 182

Betty Wilkinson Irby Collection: pages 167, 177

Carl B. King Collection: page 26

Rapp Collection: page 163

Rumbaugh-Davis Collection: page 140

Paul O. Sill Collection: page 82

Permian Historical Society Archives, University of Texas of the Permian Basin:
General Photographic Collection: pages 7, 42, 59, 175, 184
Nolan Collection: pages 5, 46, 189
Shoopman Collection: page 21

Scurry County Museum, Snyder, Texas, Delbert Hirst Collection: pages 64, 68, 84, 123, 159, 213, 237, 251

Standard Oil (New Jersey) Collection, Photographic Archives, University of Louisville: pages 93, 103, 153, 155, 243

Mrs. Joe Starkey: page 79

Texas/Dallas Collection, History and Archives Division, Dallas Public Library: page 38

Index